Advances in
CANCER
RESEARCH

Volume 104

T0348749

Clusterin, Part A

Advances in

CANCER
RESEARCH

Volume 104

Series Editors

George F. Vande Woude
Van Andel Research Institute
Grand Rapids
Michigan

George Klein
Microbiology and Tumor Biology Center
Karolinska Institute
Stockholm
Sweden

Edited by

Saverio Bettuzzi
Dipartimento di Medicina Sperimentale
Sezione di Biochimica Biochimica
Clinica e Biochimica dell'Esercizio Fisico
Via Volturno 39-43100
Parma and Istituto Nazionale
Biostrutture e Biosistemi
(I.N.B.B.) Rome, Italy

Sabina Pucci
Department of Biopathology
University of Rome
"Tor Vergata," via Montpellier
1 00-133 Rome, Italy

AMSTERDAM • BOSTON • HEIDELBERG • LONDON
NEW YORK • OXFORD • PARIS • SAN DIEGO
SAN FRANCISCO • SINGAPORE • SYDNEY • TOKYO
Academic Press is an imprint of Elsevier

ELSEVIER

Academic Press is an imprint of Elsevier
525 B Street, Suite 1900, San Diego, CA 92101-4495, USA
30 Corporate Drive, Suite 400, Burlington, MA 01803, USA
32 Jamestown Road, London, NW1 7BY, UK
Linacre House, Jordan Hill, Oxford OX2 8DP, UK
Radarweg 29, PO Box 211, 1000 AE Amsterdam, The Netherlands

First edition 2009

Notice
No responsibility is assumed by the publisher for any injury and/or damage to persons
or property as a matter of products liability, negligence or otherwise, or from any use
or operation of any methods, products, instructions or ideas contained in the material
herein. Because of rapid advances in the medical sciences, in particular, independent
verification of diagnoses and drug dosages should be made.

ISBN: 978-0-12-374772-3
ISSN: 0065-230X

For information on all Academic Press publications
visit our website at www.elsevierdirect.com

Printed and bound by CPI Group (UK) Ltd, Croydon, CR0 4YY

Transferred to Digital Print 2012

Working together to grow
libraries in developing countries

www.elsevier.com | www.bookaid.org | www.sabre.org

ELSEVIER   BOOK AID
            International   Sabre Foundation

Contents

Regulation of Clusterin Activity by Calcium

Beata Pajak and Arkadiusz Orzechowski

Nuclear CLU (nCLU) and the Fate of the Cell

Saverio Bettuzzi and Federica Rizzi

The Chaperone Action of Clusterin and Its Putative Role in Quality Control of Extracellular Protein Folding

Amy Wyatt, Justin Yerbury, Stephen Poon,
Rebecca Dabbs, and Mark Wilson

Cell Protective Functions of Secretory Clusterin (sCLU)

Gerd Klock, Markus Baiersdörfer, and Claudia Koch-Brandt

Clusterin: A Multifacet Protein at the Crossroad of Inflammation and Autoimmunity

Géraldine Falgarone and Gilles Chiocchia

Oxidative Stress in Malignant Progression: The Role of Clusterin, A Sensitive Cellular Biosensor of Free Radicals
Ioannis P. Trougakos and Efstathios S. Gonos

Color Plate Section at the end of the book

Contributors

Numbers in parentheses indicate the pages on which the authors' contributions begin.

Markus Baiersdörfer, Institute of Biochemistry, Joh.-Gutenberg University of Mainz, Becherweg 30, D-55099 Mainz, Germany (115)

Saverio Bettuzzi, Dipartimento di Medicina Sperimentale, Sezione di Biochimica, Biochimica Clinica e Biochimica dell'Esercizio Fisico, Via Volturno 39-43100 Parma and Istituto Nazionale Biostrutture e Biosistemi (I.N.B.B.), Rome, Italy (1, 9, 25, 59)

Gilles Chiocchia, AP-HP, Hôpital Ambroise Paré, Service de Rhumatologie, Boulogne-Billancourt F-92000, France; Département d'Immunologie, Inserm, U567, Paris F-75014, France; and Institut Cochin, Université Paris Descartes, CNRS (UMR 8104), Paris F-75014, France (139)

Mariangela Coletta, Dipartimento di Medicina Sperimentale, Sezione di Biochimica, Biochimica Clinica e Biochimica dell'Esercizio Fisico, Via Volturno 39-43100 Parma and Istituto Nazionale Biostrutture e Biosistemi (I.N.B.B.), Rome, Italy (9)

Rebecca Dabbs, School of Biological Sciences, University of Wollongong, Wollongong, New South Wales 2522, Australia (89)

Géraldine Falgarone, Rheumatology Department, AP-HP, Hôpital Avicenne, Bobigny F-93009, France; and EA 4222, University of Paris 13, Bobigny, France (139)

Efstathios S. Gonos, Institute of Biological Research and Biotechnology, National Hellenic Research Foundation, Athens 11635, Greece (171)

Gerd Klock, Institute of Biochemistry, Joh.-Gutenberg University of Mainz, Becherweg 30, D-55099 Mainz, Germany (115)

Claudia Koch-Brandt, Institute of Biochemistry, Joh.-Gutenberg University of Mainz, Becherweg 30, D-55099 Mainz, Germany (115)

Arkadiusz Orzechowski, Department of Physiological Sciences, Faculty of Veterinary Medicine, Warsaw University of Life Sciences (SGGW), Nowoursynowska 159, 02-776 Warsaw, Poland; and Department of Cell Ultrastructure, Mossakowski Medical Research Center, Polish Academy of Sciences, Pawinskiego 5, 02-106 Warsaw, Poland (33)

Beata Pajak, Department of Cell Ultrastructure, Mossakowski Medical Research Center, Polish Academy of Sciences, Pawinskiego 5, 02-106 Warsaw, Poland (33)

Stephen Poon, School of Biological Sciences, University of Wollongong, Wollongong, New South Wales 2522, Australia (89)

Sabina Pucci, Department of Biopathology, University of Rome "Tor Vergata," Rome, Italy (25)

Federica Rizzi, Dipartimento di Medicina Sperimentale, Sezione di Biochimica, Biochimica Clinica e Biochimica dell'Esercizio Fisico, Via Volturno 39-43100 Parma and Istituto Nazionale Biostrutture e Biosistemi (I.N.B.B.), Rome, Italy (9, 59)

Ioannis P. Trougakos, Department of Cell Biology and Biophysics, Faculty of Biology, University of Athens, Panepistimiopolis Zografou, Athens 15784, Greece (171)

Mark Wilson, School of Biological Sciences, University of Wollongong, Wollongong, New South Wales 2522, Australia (89)

Amy Wyatt, School of Biological Sciences, University of Wollongong, Wollongong, New South Wales 2522, Australia (89)

Justin Yerbury, School of Biological Sciences, University of Wollongong, Wollongong, New South Wales 2522, Australia (89)

Introduction

Saverio Bettuzzi

*Dipartimento di Medicina Sperimentale, Sezione di
Biochimica, Biochimica Clinica e Biochimica dell'Esercizio Fisico,
Via Volturno 39-43100 Parma and Istituto Nazionale
Biostrutture e Biosistemi (I.N.B.B.), Rome, Italy*

I. Introduction
References

Since the beginning, Clusterin (CLU) was revealed not as simple to study, and certainly not a single protein. The growing research interest on CLU soon produced many contributions by independent laboratories working in different systems. Thus, many different names or acronyms have been given to CLU in the early years after its discovery. Now, a general consensus recommend the name Clusterin and the abbreviation CLU. CLU was first described as a glycoprotein found nearly ubiquitous in tissues and body fluids. This early knowledge is mostly related to the secretory form of CLU (sCLU), which is exported from the cell and released in secretions acting as an extracellular chaperone. But CLU can also enter the nucleus. The detection of nCLU (nuclear CLU), which is usually associated to cell death, is now emerging as a very important event making this issue even more complex. This may explain why CLU is still often described as an "enigmatic" protein. The use of the term "enigmatic" is a clear indication that too many aspects related to the biological function(s) of CLU and its possible role in pathogenesis have been obscure, or very difficult to interpret, for long time. Contradictory findings on CLU are also present in the literature, sometimes due to technical biases or alternative interpretation of the same result. The aim of the book is ambitious: through a careful review of old data, in the light of novel information and up to date methods and hypotheses, we will try to simplify the picture for the reader and bring more light in a field still perceived to be too obscure to fully appreciate its importance and potential implementation in the clinical setting. This introduction will provide a brief general history and a critical view of the discovery of CLU with the aim to underline what is new in the field and what is now obsolete. In the rest of the book, conclusions and "take home messages" will also be provided to the reader particularly focusing on possible clinical implementations and how all this knowledge will very likely bring novelty in the fight against cancer. © 2009 Elsevier Inc.

I. INTRODUCTION

Clusterin (CLU) was first described as a glycoprotein found nearly ubiquitous in tissues and body fluids. This knowledge is mostly related to the secretory form of CLU (sCLU), which is exported from the cell and released in secretions. In the literature there is a general consensus about the apparent

Advances in CANCER RESEARCH   0065-230X/09 $35.00
Copyright 2009, Elsevier Inc. All rights reserved.   DOI: 10.1016/S0065-230X(09)04001-9

involvement of CLU in most important biological processes including sperm maturation, tissue differentiation, tissue remodeling, membrane recycling, lipid transportation, cell–cell or cell–substratum interaction, cell proliferation, cell survival, and cell death. For these reasons, CLU is generally believed to be involved in many and diverse pathological states, including neurodegeneration, ageing, and cancer (Rizzi and Bettuzzi, 2008; Rosenberg and Silkensen, 1995; Shannan *et al.*, 2006; Trougakos and Gonos, 2002; Wilson and Easterbrook-Smith, 2000).

CLU was firstly isolated from ram rete testis fluid (Fritz *et al.*, 1983). Ram rete testis fluid is known to elicit clustering of Sertoli cells (in suspension) and erythrocytes from several species. In their pioneer work, the authors showed that a heat-stable, trypsin-sensitive protein was responsible to aggregate cells. Thus, they named this extracellular protein "Clusterin," suggesting that it may play important roles in cell–cell interactions. Then, CLU was purified from the same system (Blaschuk *et al.*, 1983; Fritz *et al.*, 1983) and found to be a glycoprotein with a molecular mass of about 80 kDa and an isoelectric point of 3.6. Using reducing conditions they discovered that CLU dissociates into subunits of about 40 kDa, also showing that CLU exists both in dimeric and tetrameric forms at neutral pH and low salt concentrations. In addition, the amino acid composition of CLU was reported. Further, extracellular CLU was found to contain 4.5% glucosamine. In the same work it was suggested that Sertoli cells are the potential source of CLU, since primary cultures of rat Sertoli cells secreted CLU in the medium. One year later, different isoelectric forms of CLU were isolated (Blaschuk and Fritz, 1984). Thus, since the beginning, CLU was revealed not as simple to study, and certainly not a single protein. This basic knowledge is still true, but when addressing this issue we need to keep in mind that the most commonly reported descriptions of CLU only apply to the secreted form of CLU exported in the extracellular compartment (sCLU).

The first immunolocalization of CLU dates 1985, when Tung and Fritz raised the first monoclonal antibodies against CLU to investigate its distribution in the adult ram testis, rete testis, and excurrent ducts (Tung and Fritz, 1985). Since this first reports, the *CLU* gene was found expressed in a wide range of tissues (Choi *et al.*, 1989; de Silva *et al.*, 1990; Fischer-Colbrie *et al.*, 1984; James *et al.*, 1991), although with very different levels of expression. The research interest on CLU soon produced many contributions by independent laboratories working in different systems. Therefore, in the early years after its discovery many different names or acronyms have been given to CLU (Table I). Now, the name Clusterin and the abbreviation CLU is recommended, thanks to a general consensus among the main researchers on the field.

In humans, CLU was first described by Jenne and Tshopp in 1989 (Jenne and Tschopp, 1989) as complement cytolysis inhibitor (CLI), a component of soluble terminal complement complexes in human serum, bearing complete

Table I List of Proteins That Have Been Found Homologue of Clusterin (CLU)

Source	Species	Name	Function	Reference
Rete testes fluid	Ram	*Clusterin*	Reproduction	Blaschuk *et al.* (1983)
Adrenal medulla	Bovine	*GPIII*	Chromaffin granules	Fischer-Colbrie *et al.* (1984)
Prostate	Rat	*TRPM-2*	Apoptosis	Leger *et al.* (1987)
Prostate	Rat	*SGP-2*	Reproduction	Bettuzzi *et al.* (1989)
Neuroretinal cells	Quail	*T64*	Cell transformation	Michel *et al.* (1989)
Serum (liver)	Human	*SP-40,40*	Complement modulation	Kirszbaum *et al.* (1989)
Serum (liver)	Human	*CLI*	Complement modulation	Jenne and Tschopp (1989)
Blood	Human	*ApoJ*	Lipid transport	de Silva *et al.* (1990)
Blood	Human	*NA1/NA2*	Lipid transport	James *et al.* (1991)
Retina	Human	*K611*	Retinitis pigmentosa	Jones *et al.* (1992)

Homologues have been isolated and/or cloned by different groups working in widely divergent areas. For this reason, different names and acronyms have been originally given to the same gene/protein, today named CLU. The *CLU* gene was located in all mammalians species studied and its sequence was found highly conserved among species, while the *CLU* gene products are several proteins of different molecular size and structure also depending on the experimental system used.

identity to sulfated glycoprotein 2 (SGP2), one of the most important glycoprotein produced by Sertoli cells (Collard and Griswold, 1987). SGP2 was found, in turn, to be identical to Clusterin and suggested to play a role in sperm maturation. In the ventral prostate of the rat, CLU was first identified as testosterone repressed prostate message 2 (TRPM2) (Montpetit *et al.*, 1986). At this time, TRPM2 was described as a truncated cDNA with no correspondent protein. In an independent work published on 1989, after complete cDNA cloning, sequencing, and comparison, Bettuzzi and coworkers found that TRPM2 was instead fully homologous to full-length SGP-2 cDNA (Bettuzzi *et al.*, 1989). Only one year before, Cheng *et al.* found that SGP-2 was identical to Clusterin, the serum protein involved in aggregation of heterologous erythrocytes already described by Fritz *et al.* in 1983 (Cheng *et al.*, 1988). Finally, in 1990, an apolipoprotein named ApoJ associated with discrete subclasses of high-density lipoproteins, to which it was bound through amphiphilic helices present in its structure, was found to be the human analogue of SGP2 (de Silva *et al.*, 1990).

Therefore, the same cDNA/protein, now named CLU, was finally found to be implicated in important biological phenomena as different as erythrocyte aggregation, complement activity, lipid transport in the blood, sperm maturation, and prostate gland involution driven by androgen depletion. Historically, the first review on CLU is dated 1992 (Jenne and Tschopp, 1992).

The gene coding for CLU was mapped on chromosome 8 in the human genome (Purrello *et al.*, 1991; Slawin *et al.*, 1990; Tobe *et al.*, 1991) in position 8p12→p21 (Dietzsch *et al.*, 1992) and found to be present in single copy. The CLU gene was then found to be present in all mammalian genomes studied, with very high homology among species.

Since the beginning it appeared clear that CLU, although being a single copy gene, was actually coding for different protein products. A Western blot analysis of CLU expression in the rat ventral prostate system revealed that CLU has to be considered more as a family of protein products rather than a single protein (Fig. 1). Several antirat CLU-positive bands resolved by electrophoresis, with different molecular weights, are detectable under basal level of expression (Fig. 1, lane N) in the prostate gland. The proteomic profile of CLU is even more complex when *CLU* gene is potently induced in the regressing prostate following androgen depletion caused by surgical castration (Fig. 1, lane 1).

In spite of the growing number of papers on this issue (for instance, using Clusterin or SGP-2 or TRPM2 or ApoJ as key words, 1552 research papers and 139 reviews can be found on PubMed to date...). CLU is still often described as an "enigmatic" protein. The use of the term "enigmatic" is a clear indication that too many aspects related to the biological function(s) of

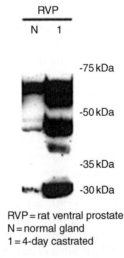

Fig. 1   Western blot analysis of CLU expression in the rat ventral prostate (RVP). N, normal prostate; 1, regressing prostate gland 4 days after androgen ablation caused by surgical castration. Several immunopositive CLU protein bands of different molecular weight are present both under basal conditions (N) and when potently induced by castration (1). During the regression of the rat prostate mainly due to massive apoptosis, extra protein bands are produced from the same gene.

CLU and its possible role in pathogenesis have been obscure, or very difficult to interpret, for long time. Contradictory findings on CLU are also present in the literature. All this do not really help our understanding of the topic.

Following a tradition which started time ago, the 5th Clusterin/Apolipo-protein J (CLU) Workshop has been held in Spetses (Greece) in June 2008. The next (6th) Workshop will be held in Parma (Italy) in 2011. The Spetses meeting brought together the most active researchers in this field. They presented their more recent data and discussed about the mechanisms through which this unique protein would be executing the fate of the cell. The debate mostly focused about whether CLU has a key role in tumor prevention or tumor promotion. The fruitful discussion started to render some consensus in the field. A scientific report on this event has been recently published (Trougakos et al., 2009). The idea to write this book has emerged during the meeting in Spestes as a great opportunity to bring together experimental data, novel ideas, scientific hypotheses, and opinions. Our hope is that the contribution of different senior researchers, focusing on the possible role of CLU in tumorigenesis, presented in this book would help the reader to understand more in this complex field.

The aim of the book is ambitious: through a careful review of old data, in the light of novel information and up to date methods and hypotheses, we will try to simplify the picture for the reader and bring more light in a field still perceived to be too obscure to fully appreciate its importance and potential implementation also in the clinical setting. After this introduction, in which a brief general history and critical view of the discovery of CLU has been provided, every chapter will be preceded by a specific introduction to better introduce the reader to the specific topics that will be further discussed. The book starts with a description of the novel knowledge and the complexity of the regulation of CLU gene expression, showing how from a single copy gene multiple transcripts and protein products can be originated in mammalian cells (Chapters "Clusterin: From one gene and two transcripts to many proteins", "The shifting balance between CLU forms during tumor progression" and "Regulation of Clusterin activity by calcium") including the nuclear form nCLU.

Because of the identification of many protein products with different molecular weights in association to intracellular processing and maturation of CLU, we will use the term "secreted CLU" (sCLU) when the most of CLU is fully maturated in the Golgi apparatus and exported from the cell as glycosylated protein in the extracellular compartment. But its secretion can be completely abolished. Under these conditions, CLU is not detectable any longer in the culture medium and the cells will get fully loaded with CLU. Now CLU can be mostly found in the cytoplasm of the cell, but also in the nucleus. The condition in which CLU is exclusively present in the cytoplasm, as well as that in which CLU is solely present in the nucleus, are very rare. The most common situation is the detection of an "intracellular" staining showing CLU both in the cytoplasm and in the nucleus in intact, fixed cells. Therefore, cytoplasm and nuclear localization may often coexist. In any case, the detection of even a small amount of CLU in the nucleus of intact cells by appropriate methods is here defined as "nuclear CLU" (nCLU). While we can track CLU with appropriate antibodies, we do not have definitive information about the structural differences between different protein products besides changes in the molecular weight as revealed by electrophoresis. Therefore, the detection of the protein forms sCLU or nCLU is here intended more as a cell functional condition rather than the identification of a structurally defined molecular form. This is why we will avoid the use of the term "isoform" in this book for CLU proteins, waiting for more definitive experimental data on the issue. The term isoform will be used instead for different transcripts coding for CLU, on which more is known because they have been identified and sequenced.

Then the discussion will continue with a critical review of what is known about sCLU and nCLU (Chapters "Nuclear CLU and the fate of the cell", "The chaperone action of Clusterin and its putative role in quality control of extracellular protein folding" and "Cell protective functions of secretory Clusterin") with a focus on their potential biological action. Then, we will address what is known today about the involvement of CLU in important physiological and pathological conditions such as inflammation and immunity (Chapter "Clusterin: A multifacet protein at the crossroad of inflammation and autoimmunity"), oxidative stress (Chapter "Oxidative stress in malignant progression: The role of Clusterin, a sensitive cellular biosensor of free radicals"), and some of the most diffuse and important kind of cancers (Chapters "CLU and prostate Cancer", "CLU and breast Cancer", "CLU and colon Cancer" and "CLU and lung Cancer" of Volume 105). The possibility that sCLU plays an important role in chemioresistance to anticancer drug is also discussed (Chapter "CLU and chemoresistance" in Volume 105), as well as the possible diverse role of CLU protein(s) and *CLU* gene expression in dependence of the local tissue context and microenvironment (Chapter "CLU and tumor microenvironment" of Volume 105). Chapter "Regulation of CLU gene expression by oncogenes and epigenetic factors: implication for tumorigenesis" of Volume 105 is the discussion arena in which several senior authors in the field will try to merge their expertise and ideas from different fields in a general consensus to provide an integrated view on CLU gene expression from a novel point of view, that is, epigenetic regulation, also discussing how *CLU* gene expression is affected by oncogenes during cell transformation.

Finally, Chapter "Conclusion and Perspectives: CLU for novel therapies and advanced diagnostic tools?" of Volume 105 will try to draw some conclusions and "take home messages", dealing with what is now obsolete in the field, which clinical implementations have been already attempted, how successful they have been in consideration of the most up to date information on this issue and how all this knowledge will very likely bring novelty in the fight against cancer.

ACKNOWLEDGMENTS

Grant sponsor: FIL 2008 and FIL 2009, University of Parma, Italy; AICR (UK) Grant No. 06–711; Istituto Nazionale Biostrutture e Biosistemi (INBB), Roma, Italy.

REFERENCES

Bettuzzi, S., *et al.* (1989). Identification of an androgen-repressed mRNA in rat ventral prostate as coding for sulphated glycoprotein 2 by cDNA cloning and sequence analysis. *Biochem. J.* **257**, 293–296.

Blaschuk, O. W., and Fritz, I. B. (1984). Isoelectric forms of clusterin isolated from ram rete testis fluid and from secretions of primary cultures of ram and rat Sertoli-cell-enriched preparations. *Can. J. Biochem. Cell Biol.* **62**, 456–461.

Blaschuk, O., *et al.* (1983). Purification and characterization of a cell-aggregating factor (clusterin), the major glycoprotein in ram rete testis fluid. *J. Biol. Chem.* **258**, 7714–7720.

Cheng, C. Y., *et al.* (1988). Rat clusterin isolated from primary Sertoli cell-enriched culture medium is sulfated glycoprotein-2 (SGP-2). *Biochem. Biophys. Res. Commun.* **155**, 398–404.

Choi, N. H., *et al.* (1989). A serum protein SP40,40 modulates the formation of membrane attack complex of complement on erythrocytes. *Mol. Immunol.* **26**, 835–840.

Collard, M. W., and Griswold, M. D. (1987). Biosynthesis and molecular cloning of sulfated glycoprotein 2 secreted by rat Sertoli cells. *Biochemistry* **26**, 3297–3303.

de Silva, H. V., *et al.* (1990). Purification and characterization of apolipoprotein J. *J. Biol. Chem.* **265**, 14292–14297.

Dietzsch, E., *et al.* (1992). Regional localization of the gene for clusterin (SP-40,40; gene symbol CLI) to human chromosome 8p12–> p21. *Cytogenet. Cell Genet.* **61**, 178–179.

Fischer-Colbrie, R., *et al.* (1984). Isolation and immunological characterization of a glycoprotein from adrenal chromaffin granules. *J. Neurochem.* **42**, 1008–1016.

Fritz, I. B., *et al.* (1983). Ram rete testis fluid contains a protein (clusterin) which influences cell-cell interactions *in vitro. Biol. Reprod.* **28**, 1173–1188.

James, R. W., *et al.* (1991). Characterization of a human high density lipoprotein-associated protein, NA1/NA2. Identity with SP-40,40, an inhibitor of complement-mediated cytolysis. *Arterioscler. Thromb.* **11**, 645–652.

Jenne, D. E., and Tschopp, J. (1989). Molecular structure and functional characterization of a human complement cytolysis inhibitor found in blood and seminal plasma: Identity to sulfated glycoprotein 2, a constituent of rat testis fluid. *Proc. Natl. Acad. Sci. USA* **86**, 7123–7127.

Jenne, D. E., and Tschopp, J. (1992). Clusterin: The intriguing guises of a widely expressed glycoprotein. *Trends Biochem. Sci.* **17**, 154–159.

Jones, S. E., *et al.* (1992). Analysis of differentially expressed genes in retinitis pigmentosa retinas. Altered expression of clusterin mRNA. *FEBS. Lett.* **300**, 279–282.

Kirszbaum, L., *et al.* (1989). Molecular cloning and characterization of the novel, human complement-associated protein, SP-40, 40: a lin between the complement and reproductive systems. *EMBO J.* **8**, 711–718.

Leger, J. G., *et al.* (1987). Characterization and cloning of androgen-repressed mRNAs from rat ventral prostate. *Biochim. Biophys. Res. Commun.* **147**, 196–203.

Michel, D., *et al.* (1989). Expression of a novel gene coding a 51.5 kD precursor protein is induced by different retroviral oncogenes in quail neuroretinal cells. *Oncogene Res.* **4**, 127–136.

Montpetit, M. L., *et al.* (1986). Androgen-repressed messages in the rat ventral prostate. *Prostate* **8**, 25–36.

Purrello, M., *et al.* (1991). The gene for SP-40,40, human homolog of rat sulfated glycoprotein 2, rat clusterin, and rat testosterone-repressed prostate message 2, maps to chromosome 8. *Genomics* **10**, 151–156.

Rizzi, F., and Bettuzzi, S. (2008). Targeting CLU in prostate cancer. *J. Physiol. Pharmacol.* **59**(Suppl. 9), 265–274.

Rosenberg, M. E., and Silkensen, J. (1995). Clusterin: Physiologic and pathophysiologic considerations. *Int. J. Biochem. Cell Biol.* **27**, 633–645.

Shannan, B., *et al.* (2006). Challenge and promise: Roles for clusterin in pathogenesis, progression and therapy of cancer. *Cell Death Differ.* **13**, 12–19.

Slawin, K., *et al.* (1990). Chromosomal assignment of the human homologue encoding SGP-2. *Biochem. Biophys. Res. Commun.* **172**, 160–164.

Tobe, T., *et al.* (1991). Assignment of a human serum glycoprotein SP-40,40 gene (CLI) to chromosome 8. *Cytogenet. Cell Genet.* **57**, 193–195.

Trougakos, I. P., and Gonos, E. S. (2002). Clusterin/apolipoprotein J in human aging and cancer. *Int. J. Biochem. Cell Biol.* **34**, 1430–1448.

Trougakos, I. P., *et al.* (2009). Advances and challenges in basic and translational research on clusterin. *Cancer Res.* **69**, 403–406.

Tung, P. S., and Fritz, I. B. (1985). Immunolocalization of clusterin in the ram testis, rete testis, and excurrent ducts. *Biol. Reprod.* **33**, 177–186.

Wilson, M. R., and Easterbrook-Smith, S. B. (2000). Clusterin is a secreted mammalian chaperone. *Trends Biochem. Sci.* **25**, 95–98.

Clusterin (CLU): From One Gene and Two Transcripts to Many Proteins

Federica Rizzi, Mariangela Coletta, and Saverio Bettuzzi

Dipartimento di Medicina Sperimentale, Sezione di Biochimica, Biochimica Clinica e Biochimica dell'Esercizio Fisico, Via Volturno 39-43100 Parma and Istituto Nazionale Biostrutture e Biosistemi (I.N.B.B.), Rome, Italy

I. Introduction
II. *CLU* Gene Organization, Promoter Region, and Transcription Products
III. CLU Protein Forms
IV. Conclusions
References

Clusterin (CLU) has kept many researchers engaged for a long time since its first discovery and characterization in the attempt to unravel its biological role in mammals. Although there is a general consensus on the fact that *CLU* is supposed to play important roles in nearly all fundamental biological phenomena and in many human diseases including cancer, after about 10 years of work CLU has been defined as an "enigmatic" protein. This sense of frustration among the researchers is originated by the fact that, despite considerable scientific production concerning *CLU*, there is still a lack of basic information about the complex regulation of its expression. The *CLU* gene is a single 9-exon gene expressed at very different levels in almost all major tissues in mammals. The gene produces at least three protein forms with different subcellular localization and diverse biological functions. The molecular mechanism of production of these protein forms remains unclear. The best known is the glycosylated mature form of CLU (sCLU), secreted with very big quantitative differences at different body sites. Hormones and growth factors are the most important regulators of *CLU* gene expression. Before 2006, it was believed that a unique transcript of about 1.9 kb was originated by transcription of the *CLU* gene. Now we know that alternative transcriptional initiation, possibly driven by two distinct promoters, may produce at least two distinct CLU mRNA isoforms differing in their unique first exon, named Isoform 1 and Isoform 2. A third transcript, named Isoform 11036, has been recently found as one of the most probable mRNA variants. Approaches like cloning, expression, and functional characterization of the different CLU protein products have generated a critical mass of information teaching us an important lesson about *CLU* gene expression regulation. Nevertheless, further studies are necessary to better understand the tissue-specific regulation of *CLU* expression and to identify the specific signals triggering the expression of different/alternative transcript isoforms and protein forms in different cell types at appropriate time. © 2009 Elsevier Inc.

Advances in CANCER RESEARCH
Copyright 2009, Elsevier Inc. All rights reserved.

0065-230X/09 $35.00
DOI: 10.1016/S0065-230X(09)04002-0

I. INTRODUCTION

The clusterin (*CLU*) gene has been independently identified in various fields without clear relationships between them (see also chapter "Introduction"). It has been associated to many biological function including tissue differentiation and remodeling, membrane recycling, lipid transportation, cell–cell or cell–substratum interaction, cell motility, cell proliferation, and cell death (Rosenberg and Silkensen, 1995; Shannan *et al.*, 2006; Trougakos and Gonos, 2002; Wilson and Easterbrook-Smith, 2000).

The multifunctional nature of *CLU* is also reflected by its complex and wide different expression pattern. CLU is a major secretion product of Sertoli cells (Collard and Griswold, 1987) and is expressed at very high and apparently constitutive levels in several cell types including motorneurons (Danik *et al.*, 1993; Michel *et al.*, 1992), dermal fibroblasts (Scaltriti *et al.*, 2004a), and some epithelia (Aronow *et al.*, 1993). In contrast, in many other cell types, clusterin gene expression is finely regulated. For instance, the induction of clusterin gene expression is tightly regulated during the cell-cycle progression (Bettuzzi *et al.*, 1999), with regard to various cases of atrophy and programmed cell death (Ahuja *et al.*, 1994; Astancolle *et al.*, 2000; Bettuzzi *et al.*, 1989) and in neurodegenerative disorders (Calero *et al.*, 2005; May and Finch, 1992). Clusterin expression is also affected by oncogenes and several reports indicate that it is downregulated upon oncogene transformation (Klock *et al.*, 1998; Kyprianou *et al.*, 1991; Lund *et al.*, 2006; Tchernitsa *et al.*, 2004; Thomas-Tikhonenko *et al.*, 2004). The relationship between *CLU* expression and the transformed phenotype has been found in many human malignant tumors (Shannan *et al.*, 2006). The wide variation in the levels of *CLU* expression in different tissues suggests that its expression is regulated in a tissue-specific manner and perhaps diverse pathways control its expression in healthy and in diseased context. A basic way to investigate the biological importance of *CLU* is to elucidate the intracellular mechanisms regulating clusterin gene induction. The human *CLU* gene was firstly characterized by Wong *et al.* (1994). Early studies about *CLU* gene products reported divergent 5′-untranslated regions (5′-UTR) in RNAs from rats (Collard and Griswold, 1987; Wong *et al.*, 1993), humans (Jenne and Tschopp, 1989; Kirszbaum *et al.*, 1989; O'Bryan *et al.*, 1990), and mice (French *et al.*, 1993; Jordan-Starck *et al.*, 1994). Therefore, the possibility of alternative first exons for clusterin transcription products in different mammalian species was already known at that time. Immediately after, Wong *et al.* (1994) characterized the human *CLU* gene and identified a unique exon I sequence corresponding to that recorded in the genomic databases at that time. While the authors correctly concluded that data obtained were not sufficient to rule out the possibility of alternative exon I usage, a simplified view prevailed, therefore until 2006 it was commonly accepted that a unique transcript of 1.9 kb was originated by transcription of the *CLU* gene (de Silva *et al.*, 1990). Now, we know that alternative transcriptional initiation, possibly driven by two distinct promoters, may produce at least two distinct CLU mRNA isoforms differing in their unique first exon. This is a mechanism well recognized in mammals, to produce distinct mRNA isoforms, with heterogeneous 5′ ends from a unique gene. Transcription from alternative promoter results in the production of mRNAs which encode for the same protein, but have a different 5′-UTR, or for distinct protein with different, or even opposite biological activities, if the first exon contains a functional ATG which originates a new main open reading frame (ORF). The complex transcriptional and translational regulation of *CLU* gene, and the existence of more than one regulatory promoter region as presented here may account for the wide different expression pattern of CLU proteins. This chapter will be, therefore, focused on the basic knowledge of *CLU* gene structure and has the main aim to provide the reader with the most updated findings on its regulation and its products.

II. *CLU* GENE ORGANIZATION, PROMOTER REGION, AND TRANSCRIPTION PRODUCTS

The clusterin (*CLU*) gene is unique in the genome and well conserved during evolution. In humans, *CLU* gene maps on chromosome 8 proximal to the lipoprotein lipase gene locus, a region that is frequently deleted in prostate cancer (8p21-p12), (Fink *et al.*, 1993). *CLU* is organized in 9 exons of variable size, ranging from 126 to 412 bp and spanning a region of 17,877 bp (Wong *et al.*, 1994). This gene organization is well conserved also in mouse and rat. The homolog mouse gene maps on chromosome 14 (Birkenmeier *et al.*, 1993) and spans a region of 12,923 bp. Exon 1 and exon 9 of the murine gene are untranslated exons. The rat homolog gene maps on chromosome 15, spanning a region of 39,536 bp.

The wide variation in the levels of CLU expression in different tissues suggests that the CLU expression is tightly regulated and very tissue specific. A sequence comparison of the 5′-flanking region of the gene revealed that the homology between the putative promoters region is confined within a very proximal domain immediately upstream the transcription start site while, the most upstream regions appear completely divergent (Michel *et al.*, 1997).

Before 2006, it was believed that *CLU* gene originates a unique transcript of 1.9 kb containing two in-frame ATG start sites, the first of which is on exon 2 (human clusterin coding sequence, GenBank accession no. M64722).

Actually, it was then found that more transcripts code for CLU, and all of them hold a third ATG, which is again in-frame with the other two, and localizes on exon 3 (Fig. 1 and Fig. 2). The proximal promoter region relative to the +1 transcription start site of the sequence M64722S revealed the presence of a conventional TATA box element as well as some potential *cis*-regulatory elements including AP-1, AP-2, and SP1 motifs (Wong *et al.*, 1994). Besides the AP1 element, a longer domain conserved among the CLU promoters in all species tested (belonging to different vertebrate classes) has been found by Michel *et al.* (1997) in A431. The authors named this 14 bp sequence as clusterin element (CLE). CLE appears to be strictly related to the heat-shock response elements (HSE), whose consensus sequence is TTCta-GAAcaTTC. The CLE sequence differs by only one base from the ideal HSE. The single mismatch is in the 5-bp repeat nGAAn central motif, which became AGAAA on the CLU promoter. This change is not believed to be able to significantly affect the binding affinity of the heat-shock factor HSF. The high base sequence conservation between distantly related species and the fact that this element is also present in the regulatory region of several other genes, like *human collagenase*, *human heme oxygenase*, and *rat c-jun*, strongly supports the possible role of CLEs as a *cis*-element of transcription. Glucocorticoid/androgen-like response elements and cyclic-AMP response elements (CREs) are also present in the 5′-flanking region of the gene.

A recent GenBank update has earmarked two transcriptional isoforms of human CLU given as RefSeq, named Isoform 1 and Isoform 2 (GenBank accession number NM_001831.2 and NM_203339.1, respectively). These two transcripts are probably originated from two alternative transcriptional initiation start sites and only produced in humans and chimpanzees. Consulting the ASAP (Alternative Splicing Annotation Project) database (Lee *et al.*, 2003) for the *CLU* UniGene cluster (Hs.75106) a third transcript, called Isoform 11036, appears to be as one of the most probable mRNA variants. The three primary transcript isoforms produced by RNA polymerase contain 9 exons, 8 introns, and a terminal 3′-UTR. All these transcripts have a unique exon 1 and share the remaining sequence from exon 2 to exon 9 (Fig. 1). In the postgenomic era, it became clear that the number of genes in eukaryotic genomes does not reflect the biological complexity of the corresponding organism. It is well known that many genes encode several variants of proteins due to the usage of alternative promoter and alternative splicing. In mammals, several genes produce distinct mRNA isoforms, with heterogeneous 5′ ends. This can generate mRNAs which encode for the same protein, but have a different 5′-UTR. In some cases, when the translation start site exists within the first exon, they encodes for distinct protein with different, alternative, or opposing biological activities. The *CLU* gene seems to be a paradigmatic example of such a complex regulation. Actually, Isoform 1 is a 5′-extended sequence end of the previously published

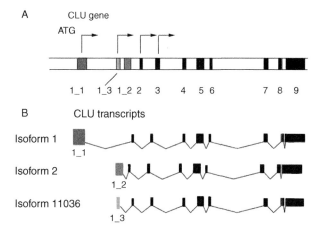

Fig. 1 Clusterin: gene and transcripts. Panel A is a schematic overview of *CLU* gene organization on human chromosome 8. Black blocks represent exons from 2 to 9, common to all transcripts, while gray blocks represent unique exon 1 of different transcriptional variants of CLU. Isoform 1 exon 1 (1_1) is indicated in dark gray, Isoform 2 exon 1 (1_2) in gray and Isoform 11036 exon 1 (1_3) in light gray. Transcriptional start sites which are in-frame in the CLU sequence are represented with black arrows. Panel B is a schematic representation of known CLU mRNA variants. (See Page 1 in Color Section at the back of the book.)

mRNA identified by the GenBank accession code M64722. Interestingly, the complete sequencing of exon 1, previously reported in a truncated form, highlighted the presence of an additional functional ATG. Therefore, this mRNA is predicted to produce a protein of 501 amino acids of a molecular weight of 57.8 kDa (Fig. 2). The same analysis allowed to suggest that the protein product should have a prevalent cytoplasmic/nuclear localization according to a computational prediction of its subcellular localization by the PSORT program (Horton *et al.*, 2007). Thus, the Isoform 1 may account for the existence of the intracellular forms of CLU escaping the secretion pathway.

At difference, Isoform 2 has an alternative untranslated exon 1. The first available ATG is located in exon 2, in a region common to the three transcripts and immediately upstream of a functional endoplasmic reticulum (ER) localization leader sequence. Using the largest ORF, this mRNA is predicted to produce a protein of 449 amino acids destined to secretion (Fig. 2). A similar prediction was made for the sequence M64722, previously published.

Two unusual features of CLU Isoform 2 mRNA are (i) a long (276 nt) 5′-UTR, corresponding to the untranslated unique exon 1, which is thought to contain extensive secondary structure; (ii) a short (57 nt) ORF within the 5′-UTR, coding for a putative small regulatory peptide. It is now known that highly structured UTRs, along with peculiar features such as richness in GC sequences, upstream ORFs and internal ribosome entry sites, significantly

MQVCSQPQRGCVREQSAINTAPPSAHNAASPGGARGHRVPLTEACKDSRIGG**M**MKTLLLFV

GLLLTWESGQVLGDQTVSDNELQE**M**SNQGSKYVNKEIQNAVNGVKQIKTLIEKTNEERKTL

LSNLEEAKKKKEDALNETRESETKLKELPGVCNETMMALWEECKPCLKQTCMKFYARVCRS

GSGLVGRQLEEFLNQSSPFYFWMNGDRIDSLLENDRQQTHMLDVMQDHFSRASSIIDELFQ

DRFFTREPQDTYHYLPFSLPHRRPHFFFPKSRIVRSLMPFSPYEPLNFHAMFQPFLEMIHE

AQQAMDIHFHSPAFQHPPTEFIREGDDDRTVCREIRHNSTGCLRMKDQCDKCREILSVDCS

TNNPSQAKLRRELDESLQVAERLTRKYNELLKSYQWKMLNTSSLLEQLNEQFNWVSRLANL

TQGEDQYYLRVTTVASHTSDSDVPSGVTEVVVKLFDSDPITVTVPVEVSRKNPKFMETVAE

KALQEYRKKHRE*STOP

Fig. 2 Translation of CLU Isoform 1 mRNA from the largest ORF. Putative Isoform 1 amino acid sequence translated from the ATG present in exon 1. Light gray highlights specific N-terminus made of 52 amino acids present only in this isoform, while dark gray indicates the leader peptide. The methionine residues coded from the three functional ATGs present in the transcriptional variant 1 are indicated in bold. The second and third methionine are common with Isoform 2. The extracellular, fully processed, and secreted sCLU is coded starting from the second methionine.

influence the rate of translation of mRNAs. This type of regulation has been extensively studied, for instance, for *S-adenosylmethionine decarboxylase (SAM-DC)*, a regulatory gene of polyamine metabolism. Deletion of virtually the entire 5′-UTR, or mutation of the internal ORF present in this region increased dramatically the protein expression, demonstrating that the 5′-UTR acts as a negative regulator (Nishimura *et al.*, 1999; Suzuki *et al.*, 1993). The protein products of other genes, like the *suppressor of cytokine signaling (SOCS-1)* (Schluter *et al.*, 2000), *retinoic acid receptor beta2 (RARbeta2)* (Peng *et al.*, 2005), and *Bcl-2* (Harigai *et al.*, 1996), are regulated at translational level and specifically repressed by a mechanism which bears many similarities to that hypothesized for SAM-DC. Such a regulation, if confirmed also for *CLU*, might account for the low level of expression that is generally found for Isoform 2.

The exon 1 of Isoform 11036 is located in the *CLU* gene sequence just between the exon 1 of Isoforms 1 and 2. This sequence is not given as a RefSeq in the GenBank database, but appears to be one of the most probable splicing products of the *CLU* gene. As previously reported for the Isoform 1, also Isoform 11036 has a functional ATG in the first exon and is supposed to produce a protein of 460 amino acids with a prevalent nuclear localization (Horton *et al.*, 2007). All transcripts hold a third ATG, which is again in-frame with the others and localizes on exon 3 (Fig. 2). Translation from this ATG would produce a shorter form of the CLU protein, which is again supposed to localize to the nuclei (Scaltriti *et al.*, 2004b).

Last but not least, an alternative messenger was found by Leskov *et al.* (2003) in MCF-7 cells, lacking exon 2. In this transcript, exon 1 of the Isoform 1 is directly joined to exon 3, therefore lacking the ATG in exon 2 and the ER-localization sequence. According to the authors, this alternative mRNA is constitutively expressed in MCF-7 breast cancer cell line, coding for a protein precursor called pnCLU, a putative nuclear pro-death form. Consulting the ASAP database, this is not one of the most probable splicing variants. Since its discovery, its existence has never been confirmed later in other cell lines or tissues by other authors (Andersen *et al.*, 2007; Cochrane *et al.*, 2007; Schepeler *et al.*, 2007). Also our attempts to amplify this messenger in prostate cancer cell lines were not successful. As shown in Fig. 3, we have been actually able to amplify the Isoform 1 and Isoform 2 of CLU mRNA from prostate cancer cell lines (PNT1a, PC3, and DU145) and from normal human embryonic fibroblasts (WI38). Isoform 1 is expressed both in normal and in tumour cell line. On the contrary, Isoform 2 is almost undetectable in all the transformed cell lines tested, being expressed at detectable levels only in normal cells. But, in the same systems, we did not find any evidence of a shorter amplicon due to the "skipping exon 2." Therefore, it should be seriously taken into consideration that this transcript is either very specific to MCF7 cells (for still unknown reasons) or an experimental artifact.

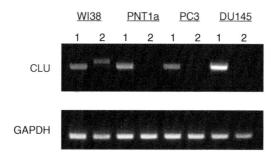

Fig. 3 Amplification of CLU transcripts by RT-PCR. CLU RT-PCR products amplified from different cell lines and resolved on a 1% agarose gel by electrophoresis are shown. Amplicons were obtained using specific forward primers able to discriminate Isoform 1 from 2. The forward primer anneals in the unique exon 1 of each transcripts, while the reverse primer anneals in exon 4. Amplicons size of variant 1 (nucleotide from 67 to 537, NM_001831.2) and variant 2 (nucleotide from 50 to 657, NM_203339.1) are 470 and 607 base pairs, respectively, as expected. The housekeeper used in the experiment and amplified in each sample for comparison is glyceraldehyde-3-phosphate dehydrogenase (GAPDH).

At difference, many evidences confirm the existence of the previously described CLU transcripts, making obsolete the idea that the human *CLU* gene produces a unique transcript. It was recently demonstrated by Andersen *et al.* (2007) that all three CLU mRNA variants are expressed in normal and transformed colorectal mucosa and cancer. Using laser microdissected (LMD) tissues and variant specific quantitative real-time RT-PCR, the authors also demonstrated a variable expression pattern of Isoform 1 in cancer epithelia respect to matched normal tissue. Interestingly, they found that Isoform 2 was always downregulated in cancer cells with respect to normal matched tissue, supporting the idea that only the level of the latter appears to be associated with colon cancer development (Andersen *et al.*, 2007). Isoform 11036 transcript was found to be expressed at very low level in normal and transformed colonic cells (Andersen *et al.*, 2007; Schepeler *et al.*, 2007).

The same research team found that signaling affecting CLU mRNA isoforms production are different. In particular, they found that the Wnt signaling pathway, via TCF1, specifically regulates the expression levels of Isoform 1 but not that of Isoform 2 (Schepeler *et al.*, 2007). Cochrane *et al.* found that the two isoforms are concomitantly expressed in prostate cancer cell line, being differentially regulated by androgens (Cochrane *et al.*, 2007). In particular, Isoform 2 is upregulated by androgens, whereas Isoform 1 is downregulated by androgens. Moreover, Isoform 2 levels do show a significant increase in levels during the progression to androgen-independent prostate cancer in LNCaP xenograft tumors (Cochrane *et al.*, 2007).

It is now clear that *CLU* encodes several mRNA variants, but still very little is known about their biological relevance and regulation. It is also unclear whether transcription of each mRNA species is driven by a unique promoter or perhaps by different promoters.

The existence of diverse human CLU transcripts is paralleled by previous findings in quail. In this system, it was demonstrated that two CLU transcripts with different, mutually exclusive, noncoding 5′ exons can be produced from the avian gene (Michel *et al.*, 1995). In contrast to the avian *CLU* gene, whose expression can be driven by two alternative promoters (Michel *et al.*, 1995), CLU expression in mammals has traditionally been thought to be controlled by one promoter only. However, the proved existence of several novel CLU mRNA variants differentially regulated by diverse signaling pathways (Cochrane *et al.*, 2007; Schepeler *et al.*, 2007), and the complexity of *CLU* regulation as illustrated by the observations that identical growth factors may elicit different *CLU* responses (Trougakos and Gonos, 2002) keep this question open. Whether transcription of each mRNA species is initiated from the same promoter or perhaps (and more reasonably) by different promoters still remains to be unravelled.

In support of the latter hypothesis, it is known since long ago that many putative transcriptional *cis*-element have been identified in the rat first intron, like the half glucocorticoid element (Wong *et al.*, 1993), an AP1 site and two transforming growth factor-β (TGF-β) inhibitory elements (Rosemblit and Chen, 1994). Recently, by androgen receptor (AR) chromatin immunoprecipitation assay it was found that the first intron of the human *CLU* gene contains the putative androgen response elements (ARE) (Cochrane *et al.*, 2007).

We have performed a simple *"in silico"* analysis to study the putative regulatory regions of *CLU* gene. We used the MatInspector software tool (Cartharius *et al.*, 2005) to find promoter regions and potential binding sites of various activator and repressor factors that bind to specific DNA regulatory sequences. Although we acknowledge that the inspection of single transcription factor binding sites (TFBS) in promoter sequences is not sufficient to fully understand gene regulation, it must be considered one of the initial and crucial events in a rationale chain of analytical steps. Interestingly, we found at least two putative promoter regions, respectively surrounding the transcription start site of Isoform 1 and Isoform 2 (Fig. 4). The first regulatory sequence (P1), is located in the 5'-flanking region of the *CLU* gene, mostly overlapping with the CLU promoter described by Wong *et al.* (1994). This promoter region has a core promoter element of about 150 nt surrounding the transcription start site of the Isoform 1 (Fig. 4). The core promoter region of P1 contains the TATA box, SP1 and SP2 sites, and the CLE element previously described. Interestingly, we have identified a second putative promoter region (P2) in the first intron of the CLU gene, immediately upstream of the transcription start site of Isoform 2 (Fig. 4). Considering that *CLU* belongs to the apolipoprotein family, it is interesting to note that intronic enhancer sequences have been shown to be involved in the expression of the

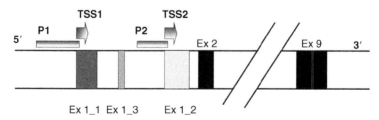

Fig. 4 Putative regulatory regions on CLU gene. The figure focuses on the organization of the *CLU* gene promoter on human chromosome 8. Black blocks represent exons from 2 to 9, common to all transcripts, while gray blocks represent unique exon 1 of different CLU transcriptional variants. TSS1 and TSS2 represent the transcription start site of Isoform 1 and Isoform 2, as confirmed by CAGE analysis. P1 and P2 represent two putative promoter regions identified by *"in silico"* analysis using the MatInspector software. P1 flanks the 5' region of the *CLU* gene, surrounds TSS1 and overlaps with the first promoter identified time ago, while P2 is located in the first intron surrounding the TSS2.

human apolipoprotein-B gene (Brooks *et al.*, 1994). The putative core element of the P2 regulatory region contains a TATA box, a CAAT box, SP1 (G/C box), but not the SP2 binding site. We did not find the CLE element, but we found a c-AMP responsive element instead and three well-conserved E-box elements, which are supposed to bind MYCN. We have also identified several STAT elements in the core promoter region and in the 5′ region of the P2 sequence. TFBS prediction programs like MatInspector may identify and suggest the presence of a regulatory site but does not provide information about its functionality *in vivo*. The binding potential of the TFBS identified "*in silico*" must ultimately be proven only by a wet-lab experiment with defined settings, particularly since potential binding sites in a promoter can be functional in a cell-, tissue-, or organ-dependent fashion, also affected by developmental stages, or nonfunctional under different conditions.

III. CLU PROTEIN FORMS

At this point, the reader may find that the situation, although still in progress, is complicated enough concerning the existing CLU transcripts. To complicate the picture further, the situation at the protein level is not any easier. It is now very clear that there are several protein forms all derived by the unique *CLU* gene, but it is unclear about how each transcript mentioned above is related to the diverse CLU protein forms observed. At least three protein forms closely related to each other have been described, with distinct subcellular and extracellular localization. Very likely, they may have different functions (Caccamo *et al.*, 2003, 2004, 2005; Moretti *et al.*, 2007; Pucci *et al.*, 2004; Scaltriti *et al.*, 2004b; Trougakos *et al.*, 2005; Yang *et al.*, 2000; Zhang *et al.*, 2006).

The secreted form of CLU (sCLU) is a 75–80 kDa glycosylated heterodimer present in almost all physiological fluids. Translation of sCLU starts from the ATG in exon 2 and produces a 427 amino acid sequence that is then targeted to the ER by an initial leader peptide. The intracytoplasmic precursor is an uncleaved, unglycosylated protein with an apparent size of 60–64 kDa by Western blot analysis before ER processing. This precursor protein is further glycosylated and proteolytically cleaved at an internal site between arg[205] and ser[206]. Two distinct monomers, with limited homology, named alpha (34–36 kDa; corresponding to residues 206–427) and beta (36–39 kDa; corresponding to residues 1–205) are linked together through five disulfide bonds (Choi-Miura *et al.*, 1992; Kirszbaum *et al.*, 1992). CLU also contains at least three potential nuclear localization signals (NLS). One is a SV-40 large T antigen-like NLS at the N-termini, residues 74–81 in humans (reference sequence: GeneBank[TM] accession number M64722);

another one is located between residues 324 and 340; a third one is located at the C-termini, residues 442–446 (Leskov *et al.*, 2003).

A shorter form of about 50 kDa targeting the cell nucleus has also been identified (nCLU). Enhanced expression of nCLU is generally associated to cell death (Caccamo *et al.*, 2003, 2004, 2005; Leskov *et al.*, 2003; Moretti *et al.*, 2007; Scaltriti *et al.*, 2004b; Yang *et al.*, 2000).

Of notice, alternative CLU protein products have never been isolated, purified, and crystallized. Therefore, their actual structure is not known. All the information we have now available on possible secondary structures of CLU forms are only predicted through the use of computational analysis programs, but never proved to be true by experimental procedures.

Secreted CLU binds a wide array of biological ligands, acting as an ATP-independent extracellular chaperon (Poon *et al.*, 2000; Wilson *et al.*, 1991). It was speculated that CLU has a flexible binding site (Bailey *et al.*, 2001) due to the combination of an intrinsically disordered region and amphipathic α-helices ordered structures. Both random coiled and molten globule regions are putatively present in the CLU protein structure. They would enable this protein to act as a biological detergent. For this reason, it has been defined as an extracellular chaperone (Bailey *et al.*, 2001). It has also been proposed that, thanks to these domains, CLU may acts as a molecular adaptor that can mediate cellular uptake and degradation of undesired macromolecules circulating in the blood, in cooperation with cognate receptors (Bartl *et al.*, 2001; Poon *et al.*, 2002a,b; Stewart *et al.*, 2007).

Recently, novel information about the pathway of degradation of CLU has filled the void of knowledge about the basic metabolism of this protein in the cell. CLU degradation occurs both by the proteasome and the lysosome pathways (Balantinou *et al.*, 2009; Rizzi *et al.*, 2009). In prostate cancer PC-3 cells CLU half-life is less than 2 h. Rapid degradation occurs through the proteasome, which may represent the main way through which prostate cancer cells avoid CLU accumulation and escape apoptotic doom (Rizzi *et al.*, 2009).

IV. CONCLUSIONS

Many experimental evidences have demonstrated the existence of diverse CLU transcripts, making obsolete the idea that the human *CLU* gene produces a unique transcript. Concerning the mechanisms driving their synthesis, recent data show that many of the previous hypotheses, based on preliminary observations, are no longer true. For instance, while alternative splicing of CLU mRNA was never confirmed, the experimental data, so far, available on the complex transcriptional regulation of *CLU* gene are now demonstrating that at least two distinct transcripts are resulting from two independent transcriptional start sites, probably driven by two distinct promoters.

The possibility that the use of alternative promoters in the *CLU* gene might result in changes in the protein motifs at the N-terminus, resulting in different subcellular localization, now represents the new challenge for future research. This topic is of particular relevance because there is a growing consensus in the literature concerning the fact that the biological function of CLU is linked to its different localization inside the cell. Therefore, determining the activity of alternative promoters under different experimental conditions and in different cell types or tissue contexts will be fundamental. The fine dissection of the genetic and epigenetic regulatory mechanism driving this phenomenon is now imperative for understanding the broad diversity of developmental processes in both normal and diseased states.

The identification of all CLU transcripts and their transcriptional regulation, together with the characterization of the structure and the intracellular localization of all translational products and their cognate functional partners, is essential to build a solid bulk of knowledge for the correct interpretation of the huge mass of data that has been produced until now. Only through this work we will be able to unravelling the biological function(s) of CLU.

CLU has been proposed as a target for cancer therapy (Chan *et al.*, 2006; Gleave *et al.*, 2002; So *et al.*, 2005). The possible link between CLU action and cancer will be discussed in chapters "CLU and prostate cancer", "CLU and breast cancer", "CLU and colon cancer", "CLU and lung cancer", "CLU and chemoresistance", "CLU and tumor microenvironment", and "Regulation of CLU gene expression by oncogenes and epigenetic factors: Implication for tumorigenesis", Vol. 105. In any case, this novel knowledge is critical and may have important clinical relevance, considering that strategies for the silencing of *CLU* gene translation through the use of antisense oligonucleotides (antisense therapy) have been already attempted to enhance apoptosis in oncologic patients. These approaches were based on the knowledge, and reference nucleotide sequences, which are now clearly obsolete. Alternative novel approaches based on upregulation of nCLU in target cells can now be hypothesized and exploited.

ACKNOWLEDGMENTS

Grant sponsor: FIL 2008 and FIL 2009, University of Parma, Italy; AICR (UK) Grant No. 06–711; Istituto Nazionale Biostrutture e Biosistemi (INBB), Roma, Italy.

REFERENCES

Ahuja, H. S., *et al.* (1994). Expression of clusterin in cell differentiation and cell death. *Biochem. Cell Biol.* 72, 523–530.

Andersen, C. L., *et al.* (2007). Clusterin expression in normal mucosa and colorectal cancer. *Mol. Cell. Proteomics* 6, 1039–1048.

Aronow, B. J., *et al.* (1993). Apolipoprotein J expression at fluid-tissue interfaces: Potential role in barrier cytoprotection. *Proc. Natl. Acad. Sci. USA* 90, 725–729.

Astancolle, S., *et al.* (2000). Increased levels of clusterin (SGP-2) mRNA and protein accompany rat ventral prostate involution following finasteride treatment. *J. Endocrinol.* 167, 197–204.

Bailey, R. W., *et al.* (2001). Clusterin, a binding protein with a molten globule-like region. *Biochemistry* 40, 11828–11840.

Balantinou, E., *et al.* (2009). Transcriptional and posttranslational regulation of clusterin by the two main cellular proteolytic pathways. *Free Radic. Biol. Med.* 46, 1267–1274.

Bartl, M. M., *et al.* (2001). Multiple receptors mediate apoJ-dependent clearance of cellular debris into nonprofessional phagocytes. *Exp. Cell Res.* 271, 130–141.

Bettuzzi, S., *et al.* (1989). Identification of an androgen-repressed mRNA in rat ventral prostate as coding for sulphated glycoprotein 2 by cDNA cloning and sequence analysis. *Biochem. J.* 257, 293–296.

Bettuzzi, S., *et al.* (1999). Clusterin (SGP-2) gene expression is cell cycle dependent in normal human dermal fibroblasts. *FEBS Lett.* 448, 297–300.

Birkenmeier, E. H., *et al.* (1993). Sulfated glycoprotein-2 (Sgp-2) maps to mouse chromosome 14. *Mamm. Genome* 4, 131–132.

Brooks, A. R., *et al.* (1994). Sequences containing the second-intron enhancer are essential for transcription of the human apolipoprotein B gene in the livers of transgenic mice. *Mol. Cell. Biol.* 14, 2243–2256.

Caccamo, A. E., *et al.* (2003). Nuclear translocation of a clusterin isoform is associated with induction of anoikis in SV40-immortalized human prostate epithelial cells. *Ann. N. Y. Acad. Sci.* 1010, 514–519.

Caccamo, A. E., *et al.* (2004). Cell detachment and apoptosis induction of immortalized human prostate epithelial cells are associated with early accumulation of a 45 kDa nuclear isoform of clusterin. *Biochem. J.* 382, 157–168.

Caccamo, A. E., *et al.* (2005). Ca^{2+} depletion induces nuclear clusterin, a novel effector of apoptosis in immortalized human prostate cells. *Cell Death Differ.* 12, 101–104.

Calero, M., *et al.* (2005). Clusterin and Alzheimer's disease. *Subcell. Biochem.* 38, 273–298.

Cartharius, K., *et al.* (2005). MatInspector and beyond: Promoter analysis based on transcription factor binding sites. *Bioinformatics* 21, 2933–2942.

Chan, J. H., *et al.* (2006). Antisense oligonucleotides: From design to therapeutic application. *Clin. Exp. Pharmacol. Physiol.* 33, 533–540.

Choi-Miura, N. H., *et al.* (1992). Identification of the disulfide bonds in human plasma protein SP-40,40 (apolipoprotein-J). *J. Biochem.* 112, 557–561.

Cochrane, D. R., *et al.* (2007). Differential regulation of clusterin and its isoforms by androgens in prostate cells. *J. Biol. Chem.* 282, 2278–2287.

Collard, M. W., and Griswold, M. D. (1987). Biosynthesis and molecular cloning of sulfated glycoprotein 2 secreted by rat Sertoli cells. *Biochemistry* 26, 3297–3303.

Danik, M., *et al.* (1993). Localization of sulfated glycoprotein-2/clusterin mRNA in the rat brain by in situ hybridization. *J. Comp. Neurol.* 334, 209–227.

de Silva, H. V., *et al.* (1990). Apolipoprotein J: Structure and tissue distribution. *Biochemistry* 29, 5380–5389.

Fink, T. M., *et al.* (1993). Human clusterin (CLI) maps to 8p21 in proximity to the lipoprotein lipase (LPL) gene. *Genomics* **16**, 526–528.

French, L. E., *et al.* (1993). Murine clusterin: Molecular cloning and mRNA localization of a gene associated with epithelial differentiation processes during embryogenesis. *J. Cell Biol.* **122**, 1119–1130.

Gleave, M., *et al.* (2002). Antisense therapy: Current status in prostate cancer and other malignancies. *Cancer Metastasis Rev.* **21**, 79–92.

Harigai, M., *et al.* (1996). A *cis*-acting element in the BCL-2 gene controls expression through translational mechanisms. *Oncogene* **12**, 1369–1374.

Horton, P., *et al.* (2007). WoLF PSORT: Protein localization predictor. *Nucleic Acids Res.* **35**, W585–W587.

Jenne, D. E., and Tschopp, J. (1989). Molecular structure and functional characterization of a human complement cytolysis inhibitor found in blood and seminal plasma: Identity to sulfated glycoprotein 2, a constituent of rat testis fluid. *Proc. Natl. Acad. Sci. USA* **86**, 7123–7127.

Jordan-Starck, T. C., *et al.* (1994). Mouse apolipoprotein J: Characterization of a gene implicated in atherosclerosis. *J. Lipid Res.* **35**, 194–210.

Kirszbaum, L., *et al.* (1989). Molecular cloning and characterization of the novel, human complement-associated protein, SP-40,40: A link between the complement and reproductive systems. *Embo J.* **8**, 711–718.

Kirszbaum, L., *et al.* (1992). SP-40,40, a protein involved in the control of the complement pathway, possesses a unique array of disulphide bridges. *FEBS Lett.* **297**, 70–76.

Klock, G., *et al.* (1998). Differential regulation of the clusterin gene by Ha-ras and c-myc oncogenes and during apoptosis. *J. Cell Physiol.* **177**, 593–605.

Kyprianou, N., *et al.* (1991). Programmed cell death during regression of the MCF-7 human breast cancer following estrogen ablation. *Cancer Res.* **51**, 162–166.

Lee, C., *et al.* (2003). ASAP: The Alternative Splicing Annotation Project. *Nucleic Acids Res.* **31**, 101–105.

Leskov, K. S., *et al.* (2003). Synthesis and functional analyses of nuclear clusterin, a cell death protein. *J. Biol. Chem.* **278**, 11590–11600.

Lund, P., *et al.* (2006). Oncogenic HRAS suppresses clusterin expression through promoter hypermethylation. *Oncogene* **25**, 4890–4903.

May, P. C., and Finch, C. E. (1992). Sulfated glycoprotein 2: New relationships of this multifunctional protein to neurodegeneration. *Trends Neurosci.* **15**, 391–396.

Michel, D., *et al.* (1992). Possible functions of a new genetic marker in central nervous system: The sulfated glycoprotein-2 (SGP-2). *Synapse* **11**, 105–111.

Michel, D., *et al.* (1995). The expression of the avian clusterin gene can be driven by two alternative promoters with distinct regulatory elements. *Eur. J. Biochem.* **229**, 215–223.

Michel, D., *et al.* (1997). Stress-induced transcription of the clusterin/apoJ gene. *Biochem. J.* **328**(Pt. 1), 45–50.

Moretti, R. M., *et al.* (2007). Clusterin isoforms differentially affect growth and motility of prostate cells: Possible implications in prostate tumorigenesis. *Cancer Res.* **67**, 10325–10333.

Nishimura, K., *et al.* (1999). Gene structure and chromosomal localization of mouse S-adenosylmethionine decarboxylase. *Gene* **238**, 343–350.

O'Bryan, M. K., *et al.* (1990). Human seminal clusterin (SP-40,40). Isolation and characterization. *J. Clin. Invest.* **85**, 1477–1486.

Peng, X., *et al.* (2005). Identification of novel RARbeta2 transcript variants with short 5′-UTRs in normal and cancerous breast epithelial cells. *Oncogene* **24**, 1296–1301.

Poon, S., *et al.* (2000). Clusterin is an ATP-independent chaperone with very broad substrate specificity that stabilizes stressed proteins in a folding-competent state. *Biochemistry* **39**, 15953–15960.

Poon, S., *et al.* (2002a). Mildly acidic pH activates the extracellular molecular chaperone clusterin. *J. Biol. Chem.* **277**, 39532–39540.

Poon, S., *et al.* (2002b). Clusterin is an extracellular chaperone that specifically interacts with slowly aggregating proteins on their off-folding pathway. *FEBS Lett.* **513**, 259–266.

Pucci, S., *et al.* (2004). Modulation of different clusterin isoforms in human colon tumorigenesis. *Oncogene* **23**, 2298–2304.

Rizzi, F., *et al.* (2009). Clusterin is a short half-life, poly-ubiquitinated protein, which controls the fate of prostate cancer cells. *J. Cell. Physiol.* **219**, 314–323.

Rosemblit, N., and Chen, C. L. (1994). Regulators for the rat clusterin gene: DNA methylation and *cis*-acting regulatory elements. *J. Mol. Endocrinol.* **13**, 69–76.

Rosenberg, M. E., and Silkensen, J. (1995). Clusterin: Physiologic and pathophysiologic considerations. *Int. J. Biochem. Cell Biol.* **27**, 633–645.

Scaltriti, M., *et al.* (2004a). Clusterin (SGP-2, ApoJ) expression is downregulated in low- and high-grade human prostate cancer. *Int. J. Cancer* **108**, 23–30.

Scaltriti, M., *et al.* (2004b). Intracellular clusterin induces G2-M phase arrest and cell death in PC-3 prostate cancer cells1. *Cancer Res.* **64**, 6174–6182.

Schepeler, T., *et al.* (2007). Clusterin expression can be modulated by changes in TCF1-mediated Wnt signaling. *J. Mol. Signal.* **2**, 6.

Schluter, G., *et al.* (2000). Evidence for translational repression of the SOCS-1 major open reading frame by an upstream open reading frame. *Biochem. Biophys. Res. Commun.* **268**, 255–261.

Shannan, B., *et al.* (2006). Challenge and promise: Roles for clusterin in pathogenesis, progression and therapy of cancer. *Cell Death Differ.* **13**, 12–19.

So, A., *et al.* (2005). Antisense oligonucleotide therapy in the management of bladder cancer. *Curr. Opin. Urol.* **15**, 320–327.

Stewart, E. M., *et al.* (2007). Effects of glycosylation on the structure and function of the extracellular chaperone clusterin. *Biochemistry* **46**, 1412–1422.

Suzuki, T., *et al.* (1993). Polyamine regulation of *S*-adenosylmethionine decarboxylase synthesis through the 5′-untranslated region of its mRNA. *Biochem. Biophys. Res. Commun.* **192**, 627–634.

Tchernitsa, O. I., *et al.* (2004). Transcriptional basis of KRAS oncogene-mediated cellular transformation in ovarian epithelial cells. *Oncogene* **23**, 4536–4555.

Thomas-Tikhonenko, A., *et al.* (2004). Myc-transformed epithelial cells down-regulate clusterin, which inhibits their growth *in vitro* and carcinogenesis *in vivo*. *Cancer Res.* **64**, 3126–3136.

Trougakos, I. P., and Gonos, E. S. (2002). Clusterin/apolipoprotein J in human aging and cancer. *Int. J. Biochem. Cell Biol.* **34**, 1430–1448.

Trougakos, I. P., *et al.* (2005). Differential effects of clusterin/apolipoprotein J on cellular growth and survival. *Free Radic. Biol. Med.* **38**, 436–449.

Wilson, M. R., and Easterbrook-Smith, S. B. (2000). Clusterin is a secreted mammalian chaperone. *Trends Biochem. Sci.* **25**, 95–98.

Wilson, M. R., *et al.* (1991). Clusterin enhances the formation of insoluble immune complexes. *Biochem. Biophys. Res. Commun.* **177**, 985–990.

Wong, P., *et al.* (1993). Genomic organization and expression of the rat TRPM-2 (clusterin) gene, a gene implicated in apoptosis. *J. Biol. Chem.* **268**, 5021–5031.

Wong, P., *et al.* (1994). Molecular characterization of human TRPM-2/clusterin, a gene associated with sperm maturation, apoptosis and neurodegeneration. *Eur. J. Biochem.* **221**, 917–925.

Yang, C. R., *et al.* (2000). Nuclear clusterin/XIP8, an x-ray-induced Ku70-binding protein that signals cell death. *Proc. Natl. Acad. Sci. USA* **97**, 5907–5912.

Zhang, Q., *et al.* (2006). The leader sequence triggers and enhances several functions of clusterin and is instrumental in the progression of human prostate cancer *in vivo* and *in vitro*. *BJU Int.* **98**, 452–460.

The Shifting Balance Between CLU Forms During Tumor Progression

Sabina Pucci* and Saverio Bettuzzi[†]

*Department of Biopathology, University of Rome
"Tor Vergata," Rome, Italy
[†]Dipartimento di Medicina Sperimentale, Sezione di Biochimica, Biochimica
Clinica e Biochimica dell'Esercizio Fisico, Via Volturno 39-43100 Parma and
Istituto Nazionale Biostrutture e Biosistemi (I.N.B.B.), Rome, Italy

Cell transformation is strictly linked to important metabolic changes which are instrumental for initial survival of cancer cells and subsequent spreading of disease. Early (i.e., anaerobic glycolysis) and late metabolic changes (i.e., fatty acid metabolism) are required for progression and clinical emergence of cancer. Besides well-known tumor suppressors and oncogenes, several metabolic genes have been found implicated in this multistep process, among which are fatty acid synthase (FASN) and carnitine palmitoyl transferase I (CPT I). An intriguing link between these metabolic shifts and a change in the balance between nuclear and secreted forms of CLU (nCLU/sCLU) has been suggested. The shifting balance between CLU forms during tumor progression, by affecting the fate of the cell, seems to be strongly influenced by the metabolic shift occurring in the different steps of tumor progression. © 2009 Elsevier Inc.

I. INTRODUCTION

In tumor progression, the overall metabolic demand of neoplastic cells is significantly higher than most other tissues and the cancer cell depends more on glycolysis, even in the presence of available oxygen. Enhanced "aerobic glycolysis" in cancer cells is known as the Warburg effect (Shaw, 2006). Though controversial over the years, a molecular basis for the Warburg effect is emerging from genetic and pharmacological studies which demonstrate that specific oncogene and tumor suppressor mutations or dysregulation directly control glycolysis and oxidative phosphorylation. The combination of these mutations and the hypoxic conditions in many tumor types is likely to synergize and to control the CLU expression depending on the overall

Advances in CANCER RESEARCH
Copyright 2009, Elsevier Inc. All rights reserved.

0065-230X/09 $35.00
DOI: 10.1016/S0065-230X(09)04003-2

metabolic state of individual tumors. Therefore, changes in CLU forms expression could be strongly influenced by the metabolic shift occurring in the different steps of tumor progression.

II. SHIFTING OF CELL METABOLISM IN TUMORIGENESIS

Mutated cancer cells could benefit from glycolysis in many ways, in fact glycolysis generates more energy more quickly than in normal cells: unlike to the normal cell, the neoplastic cell typically utilizes nutrients in an "energy-independent" manner, as a consequence it does not need to break down amino acids and fatty acids to generate energy and it can utilize them to build proteins and lipids necessary for growth. The proliferative advantage of glycolysis is further demonstrated by *in vitro* experiments showing that the switch back to oxidative phosphorylation from glycolysis is accompanied by a decrease in cell growth and tumorigenicity. The long-chain fatty acids accumulated can be toxic and also induce cell death. Hence, we can speculate that the shifting of cell metabolism controls tumor cell growth and apoptosis. The nutrient deprivation and energy stress in nonmutated cell activate a program that inhibits cell-cycle progression and biosynthetic process through LKB1-activated AMPK. On the other hand, when cell lacks tumor suppressors (such as TSC2, LKB1, p53) or carries active onco-genes (Ras, Akt, Her2) mTOR and HIF1 pathways are activated, increasing cell growth and protein and lipid synthesis through fatty acid synthase (FASN) overexpression.

FASN catalyzes the synthesis of palmitate from the condensation of malonyl-CoA and acetyl-CoA and plays an important role in energy homeo-stasis by converting excess carbon intake into fatty acids for storage. In normal cells, FASN is expressed at low levels due to the presence of dietary lipids. In contrast, neoplastic cells can either use endogenously synthetic fatty acids to satisfy their metabolic necessities and to support membrane synthesis. Menendez *et al.* (2005a,b) show that the extracellular acidosis present in the microenvironment of solid tumors can work in an epigenetic fashion by upregulating the transcriptional expression of FASN gene in breast cancer cells. Moreover, The PI3K/AKT signaling pathway has also been implicated in the regulation of FASN expression. A positive feedback loop has been proposed between AKT activation and FASN expression (Wang *et al.*, 2005). In fact, the well-characterized oncogene Her-2/Neu and its downstream effector PI3K have a stimulatory effect on FASN gene. It has been shown that FASN expression is markedly increased in several human malignancies, notably breast and prostate cancer, and its overexpression in tumor tissues from patients with colon, breast and prostate carcinomas as well as melanoma and gastrointestinal stromal tumors has been associated with a poor prognosis (Pizer *et al.*, 2001). In addition, one-fourth of human prostate cancers have genomic amplification of FASN (Rossi *et al.*, 2003).

It has been recently demonstrated that FASN is overexpressed also in prostate intraepithelial neoplasia (PIN) compared with adjacent normal tissue, suggesting that it plays a role in the initial phases of prostate tumorigenesis, and in metastatic prostate cancer, suggesting that it may function as a mediator of biological aggressiveness. Importantly, Rossi (Rossi *et al.*, 2003) demonstrated that FASN-overexpressing prostate cancers display a characteristic gene expression signature, indicating a particular transcription pattern related to its activity. Its role in cancer supports the hypothesis that FASN is a metabolic enzyme and candidate oncogene in cancer (Migita *et al.*, 2009).

Overall these data support the role of FASN as a novel "*metabolic oncogene*" in cancer cells (Menendez *et al.*, 2005a,b). From a functional standpoint, pharmacological approaches to decrease expression of FASN have been shown to result in growth inhibition of various tumor cell lines, including those derived from prostate cancer and/or prostate cancer tumor xenografts *in vivo* (Migita *et al.*, 2009). In addition, functional interference, mostly by RNA interference, has been shown to result in G1 arrest and/or induction of apoptosis. FASN-specific inhibitors such as mycotoxin cerulenin and its derivative C75, the beta-lactone orlistat, the green tea polyphenol EGCG, and the novel and potent inhibitors of FASN derived from green tea catechins (GTC) have been reported to induce programmed cell death in cancer cells (Zhang *et al.*, 2008).

III. SHIFTING OF CLU FORMS DURING TUMOR PROGRESSION

Pucci S. and her group have found that the increasing endogenously synthesized fatty acids together with increasing levels of IL-6, also induced by Her2/Neu signaling and sustained in breast cancer by autocrine loop, induce high level of prosurvival sCLU in human breast carcinomas and *in vitro* breast cancer cell lines with or without Her2 gene amplification (SKBR3, MCF7 cells). The expression levels of FASN would also be in breast cancer cells, an indicator of Her2 transduction activity. The amplification of Her2 is present in 25% of breast cancer but it could be commonly present in various epithelial tumors that overexpress FASN. FASN-derived phospholipids, the end product of nearly 85% of all lipids synthesized *de novo* by

FASN in tumor cells, have been observed to end up in lipid rafts in plasma membranes (Menendez *et al.*, 2005a,b). In tumor progression, general alterations in the lipid compositions of the cellular and mitochondrial membranes may confer a selective growth advantage to aberrant cells that display increased FASN activity by inhibition of apoptosis. A strong increase of sCLU in the cytoplasm of tumoral tissues was correlated with FASN protein levels. sCLU, which binds hydrophobic macromolecules, would act in this context to "clear" potentially harmful cellular components, enhancing survival of cancerous cell.

Treatment *in vitro* with hydrocortisone strikingly induces an increase of sCLU protein. A strong increase of sCLU was found in FASN overexpressing breast tumors, which was detected in the cytoplasm bound to Ku70-Bax complex, inhibiting Bax-dependent cell death activation in breast cancer cells. These interactions among Ku–CLU–Bax represent one of cell death escaping mechanism common in colon cancers as previously published by Pucci *et al.* (2009). FASN inhibition by cerulenin induces an increased expression and the accumulation of nCLU in the nuclei of breast cancer cells, more evident in the Her2-amplified SKBR3 cells, favoring the proapoptotic pathway reverting the neoplastic apoptosis resistance.

Moreover, FASN inhibitors induced an increase of Ku70 acetylation and subsequent sCLU–KU70 release from Bax, sterically inhibited in tumors by these interactions. Therefore, the apoptotic processes induced through the inhibition of the oncogene FASN involve BAX heterodimerization and migration to mitochondria and the accumulation of nCLU in the nuclei. These observations suggest that FASN overexpression may protect prostate epithelial cells from apoptosis, while inhibition of FASN expression could induce the proapoptotic form of CLU in cancer cells.

All together, data suggest a link among tumor progression, cell metabolic shift, sCLU/nCLU balance, and cell fate in neoplastic cells. These observations on breast and colon cancer cells are in agreement with data obtained by Bettuzzi and his group in TRAMP mouse prostate carcinoma model using GTC, known to inhibit the activity of FASN (Scaltriti *et al.*, 2006).

They reported the effect of the GTC, known to display chemopreventive effects in many cancer models, including transgenic adenocarcinoma mouse prostate (TRAMP) mice that spontaneously develop prostate cancer (CaP) on nCLU production. In particular, CLU expression was detectable at basal levels in young TRAMP mice, being potently downregulated together with caspase-9 during onset and progression of CaP. At difference, in TRAMP mice treated with the FASN inhibitors GTC, tumor progression was chemoprevented and CLU mRNA and protein progressively accumulated in the prostate gland, while caspase-9 expression also returned to basal levels. Massive nuclear staining with anti-CLU antibodies of prostate cells (nCLU) was demonstrated at early stages of chemoprevention treatment with GTC. These data suggested CLU as tumor suppressor in prostate cancer, possibly mediating the anticancer effect of GTC (Caporali *et al.* 2004). Interestingly, in a further study (Scaltriti *et al.*, 2006), molecular classification of GTC-sensitive versus GTC-resistant prostate cancer was successfully attempted in the TRAMP mice model by quantitative real-time PCR gene profiling. A set of eight informative genes previously identified (Bettuzzi *et al.*, 2003) was used for molecular classification. Linear discriminant analysis was performed to discriminate four mice classes: wild type, TRAMP spontaneously developing CaP, GTC-sensitive (chemo-prevented) TRAMP, and GTC-resistant TRAMP in which administration of GTC failed to prevent CaP progression. In this study, different combinations of two genes at a time extracted from the whole set of eight genes (with a total of 28 different possible combinations) were taken into consideration. Among these combinations, best performing one was CLU-GAPDH, a well-known enzyme of the glycolysis pathway, with a 0% misclassification ratio over the four classes studied. As described previously, this result connecting dysregulation of CLU expression and glycolysis in cancer cells can be interpreted as due to the metabolic shift occurring in the different steps of tumor progression as hypothesized by the Warburg effect.

Besides the role of FASN in tumor development and progression, it is to note the behavior of another "player" of cell metabolism, carnitine palmitoyl transferase I (CPT I) and its new role in the regulation of DNA acetylation and CLU expression in the tumoral context. CPT I in normal cells resides at the outer mitochondrial membrane and it serves to transport long-chain fatty acids into mitochondria for beta-oxidation. Physiologically, the overexpression of FASN downmodulates CPT I activity, reducing oxidation of newly synthesized fatty acids suggesting a reciprocal regulation that could become aberrant in neoplastic cells.

Two isoforms of CPT1 have been characterized, known as L-CPT1 (CPT1A) and M-CPT1 (CPT1B) in liver (L-) and muscle (M-), respectively, where the expression of each was initially described, showing overlapping tissue-specific expression. While CPT1B is expressed in skeletal muscle, heart, testis, and adipose tissue, CPT1A has a more widespread distribution (Weis et al, 1994). CPT I isoform switching has been shown to take place in development when programmed cell death is enhanced.

The precursors of sphingolipids like palmitoyl-CoA are subject to removal from the cytoplasm by CPT I, thus CPT I activity may limit *de novo* synthesis of sphingolipids. Therefore, a feedback loop transregulate the activities of FASN and CPTI. The long-chain fatty acids such as palmitate and stearate can cause programmed cell death correlated with *de novo* synthesis of ceramide (Paumen *et al.*, 1997). It was demonstrated that CPT I interacts with Bcl-2 protein, that regulates programmed cell death in several systems and it is also expressed at the outer mitochondrial membrane (Reed, 1994).

Bcl-2 binding to CPT I may modulate sphingolipid metabolism in a yet to be defined way and it would control a cell death-specific activity of CPT I at the mitochondrial membrane. In addition, the carnitine system (which comprises carnitine, CPT I, carnitine acetyl transferase, and carnitine translocase) plays an important role in the cell-trafficking of short-chain fatty acids such as acetyl-CoA and works to maintain the acetyl-CoA/CoA ratio (Bremer, 1997).

Therefore, CPT I activity would confer a protective effect on normal cell viability, by the clearance of long-chain fatty acyl-CoA from the cytoplasm.

Mazzarelli et al. (2007) recently observed that the CPT I was significantly decreased in the mitochondria, and it strikingly localized in the nuclei of tumoral tissues (colon, breast, liver, ovary). At this purpose, in vitro experiments using epithelial neoplastic (MCF-7, Caco-2, HepG2) cells and nonneoplastic cell lines (MCF-12F) confirmed a nuclear localization of CPT1 protein exclusively in neoplastic cells. Moreover, histone deacetylase (HDAC) activity showed significantly higher levels in nuclear extracts from neoplastic than from control cells. HDAC1 and CPT1 proteins were coimmunoprecipitated in nuclear extracts from MCF-7 cells. The treatment with HDAC inhibitors such as trichostatin A and butyrate significantly decreased nuclear expression of CPT1 and its binding to HDAC1. The existence of CPT1A mRNA transcript variant 2 in MCF-7, besides the classic isoform 1 was also characterized. Mazzarelli et al. observed that CPT I in the nucleus could be implicated in the epigenetic regulation of gene transcription, a relevant process to control tumor growth. In fact, the peculiar localization of CPT1 in the nuclei of human carcinomas and the disclosed functional link between nuclear CPT1 and HDAC1 propose a new role of CPT1 in the histonic acetylation level of tumors. In neoplastic cells FASN overexpression inhibited β-oxidation, thus CPT I protein possibly could move to the nucleus and modulate the acetyl moieties at histone level. Moreover, the silencing of CPT1A nuclear expression by small-interfering RNAs is a sufficient condition to induce apoptosis in MCF-7 breast cancer cells where FASN and sCLU are concomitantly overexpressed. The apoptosis triggered by RNA interference correlates with reduction of HDAC activity and hyperacetylation of histone- and nonhistone proteins, involved in cancer-relevant death pathways. Moreover, the CPT1A knockdown induces downstream effects on proapoptotic genes (upregulation) and invasion- and metastasis-related genes (downmodulation), as shown by microarray analysis. Downstream effects induced by histone hyperacetylation were the upregulation of proapoptotic transcription (BAD, CASP9, COL18A1) and the downmodulation of invasion and metastasis-related genes (TIMP-1, PDGF-A, SERPINB2). Focusing on the cell-death pathway, CPT1A silencing induced cell death modulating Ku70 acetylation state and affecting sCLU–Ku70–Bax interactions and inducing nCLU accumulation in the nucleus.

IV. CONCLUDING REMARKS

The results provided above bring important evidences of existing mechanisms linking tumor progression to the metabolism-dependent epigenetic control to cell-death escape. In particular, FASN and CLU form production are specifically regulated in cancer cells which merely retain CPT1A nuclear localization. The possibility to revert these aberrant interactions by RNA targeting could offer new strategies for innovative cancer therapies.

ACKNOWLEDGMENTS

S.B. grant sponsors:

– FIL 2008 and FIL 2009, University of Parma, Italy; AICR (UK) Grant No. 06-711; Istituto Nazionale Biostrutture e Biosistemi (INBB), Roma, Italy.
– Alleanza Contro ilCancro (ACC)-Istituto superiore di Sanità (ISS) Art.3 DM 21 luglio 2006-Programma Straordinario di Ricerca Oncologica 2006-Programma 3; Progetto ordinario IRCCS.

REFERENCES

Bettuzzi, S., Scaltriti, M., Caporali, A., Brausi, M., D'Arca, D., Astancolle, S., Davalli, P., and Corti, A. (2003). Successful prediction of prostate cancer recurrence by gene profiling in combination with clinical data: A 5 years follow-up study. *Cancer Res.* **63**, 3469–3472.

Bremer, J. (1997). The role of carnitine in cell metabolism. *In* "Carnitine Today," (C. De Simone and G. Famularo, Eds.), pp. 1–37. Springer-Verlag, Heidelberg.

Caporali, A., Davalli, P., Astancolle, S., D'Arca, D., Brausi, M., Bettuzzi, S., and Corti, A. (2004). The chemopreventive action of catechins in the TRAMP mouse model of prostate carcinogenesis is accompanied by clusterin over-expression. *Carcinogenesis* **25**, 2217–2224.

Mazzarelli, P., Pucci, S., Bonanno, E., Sesti, F., Calvani, M., and Spagnoli, L. G. (2007). Carnitine palmitoyltransferase I in human carcinomas: A novel role in histone deacetylation? *Cancer Biol. Ther.* **6**, 1606–1613.

Menendez, J. A., Vellon, L., and Lupu, R. (2005a). Targeting fatty acid synthase-driven lipid rafts: A novel strategy to overcome trastuzumab resistance in breast cancer cells. *Med. Hypotheses* **64**, 997–1001.

Menendez, J. A., Decker, J. P., and Lupu, R. (2005b). In support of fatty acid synthase (FAS) as a metabolic oncogene: Extracellular acidosis acts in an epigenetic fashion activating FAS gene expression in cancer cells. *J. Cell. Biochem.* **94**(1), 1–4.

Migita, T., Ruiz, S., Fornari, A., Fiorentino, M., Priolo, C., Zadra, G., Inazuka, F., Grisanzio, C., Palescandolo, E., Shin, E., Fiore, C., Xie, W., *et al.* (2009). Fatty acid synthase: A metabolic enzyme and candidate oncogene in prostate cancer. *J. Natl Cancer Inst.* **101**(7), 519–532.

Paumen, M. B., Ishida, Y., Muramatsu, M., Yamamoto, M., and Honjo, T. (1997). Inhibition of carnitine palmitoyltransferase I augments sphingolipid synthesis and palmitate-induced apoptosis. *J. Biol. Chem.* **272**(6), 3324–3329.

Pizer, E. S., Pflug, B. R., Bova, G. S., Han, W. F., Udan, M. S., and Nelson, J. B. (2001). Increased fatty acid synthase as a therapeutic target in androgen-independent prostate cancer progression. *Prostate* 47(2), 102–110.

Pucci, S., Mazzarelli, P., Sesti, F., Boothmann, D. A., and Spagnoli, L. G. (2009). Interleukin-6 affects cell death escaping mechanisms acting on Bax–Ku70–Clusterin interactions in human colon cancer progression. *Cell Cycle* 8(3), 473–481.

Reed, J. C. (1994). Bcl-2 and the regulation of programmed cell death. *J. Cell Biol.* 124(1–2), 1–6.

Rossi, S., Graner, E., Febbo, P., Weinstein, L., Bhattacharya, N., Onody, T., Bubley, G., Balk, S., and Loda, M. (2003). Fatty acid synthase expression defines distinct molecular signatures in prostate cancer. *Mol. Cancer Res.* 1(10), 707–715.

Scaltriti, M., Belloni, L., Caporali, A., Davalli, P., Remondini, D., Rizzi, F., Astancolle, S., Corti, A., and Bettuzzi, S. (2006). Molecular classification of green tea catechins-sensitive and -resistant prostate cancer in the TRAMP mice modelby quantitative real-time PCR gene profiling. *Carcinogenesis* 27, 1047–1053.

Shaw, R. J. (2006). Glucose metabolism and cancer. *Curr. Opin. Cell Biol.* 18, 598–608.

Wang, H. Q., Altomare, D. A., Skele, K. L., Poulikakos, P. I., Kuhajda, F. P., Di Cristofano, A., and Testa, J. R. (2005). Positive feedback regulation between AKT activation and fatty acid synthase expression in ovarian carcinoma cells. *Oncogene* 24(22), 3574–3582.

Weis, B. C., Esser, V., Foster, D. W., and McGarry, J. D. (1994). Rat heart expresses two forms of mitochondrial carnitine palmitoyltransferase I. The minor component is identical to the liver enzyme. *J. Biol. Chem.* 269, 18712–18715.

Zhang, S. Y., Ma, X. F., Zheng, C. G., Wang, Y., Cao, X. L., and Tian, W. X. (2008). Novel and potent inhibitors of fatty acid synthase derived from catechins and their inhibition on MCF-7 cell. *J. Enzyme Inhib. Med. Chem.* 1, 1.

Regulation of Clusterin Activity by Calcium

Beata Pajak* and Arkadiusz Orzechowski*,†

*Department of Cell Ultrastructure, Mossakowski Medical Research Center,
Polish Academy of Sciences, Pawinskiego 5, 02-106 Warsaw, Poland
†Department of Physiological Sciences, Faculty of Veterinary
Medicine, Warsaw University of Life Sciences (SGGW),
Nowoursynowska 159, 02-776 Warsaw, Poland

In this chapter, the attention is put on Ca^{2+} effect on Clusterin (CLU) activity. We showed that two CLU forms (secreted and nuclear) are differently regulated by Ca^{2+} and that Ca^{2+} fluxes affect *CLU* gene expression. A secretory form (sCLU) protects cell viability whereas nuclear form (nCLU) is proapoptotic. Based on available data we suggest, that different CLU forms play opposite roles, depending on intracellular Ca^{2+} concentration, time-course of Ca^{2+} current, intracellular Ca^{2+} compartmentalization, and final Ca^{2+} targets. Discussion will be motivated on how CLU acts on cell in response to Ca^{2+} waves. The impact of Ca^{2+} on *CLU* gene activity and transcription, posttranscriptional modifications, translation of CLU mRNA, and posttranslational changes as well as biological effects of CLU will be discussed. We will also examine how Ca^{2+} signal and Ca^{2+}-dependent proteins are attributable to changes in CLU characteristics. Some elucidation of *CLU* gene activity, CLU protein formation, maturation, secretion, and intracellular translocations in response to Ca^{2+} is presented. In response to cell stress (i.e., DNA damage) *CLU* gene is activated. We assume that commonly upregulated mRNA for nCLU versus sCLU and vice versa are dependent on Ca^{2+} accessibility and its intracellular distribution. It looks as if at low intracellular Ca^{2+} the delay in cell cycle allows more time for DNA repair; otherwise, cells undergo nCLU-dependent apoptosis. If cells are about to survive, intrinsic apoptosis is abrogated by sCLU interacting with activated Bax. In conclusion, a narrow range of intracellular Ca^{2+} concentrations is responsible for the decision whether nCLU is mobilized (apoptosis)

Advances in CANCER RESEARCH
Copyright 2009, Elsevier Inc. All rights reserved.

0065-230X/09 $35.00
DOI: 10.1016/S0065-230X(09)04004-4

or sCLU is appointed to improve survival. Since the discovery of CLU, a huge research progress has been done. Nonetheless we feel that much work is left ahead before remaining uncertainties related to Ca^{2+} signal and the respective roles of CLU proteins are unraveled. © 2009 Elsevier Inc.

I. INTRODUCTION

In modern molecular biology, if any concern is put on functions played by a particular protein, the role of calcium signal should be addressed. Particularly in figuring out the triggering mechanisms that initiate secretion, intracellular translocations, or protein interactions. CLU protein is among those proteins that behave in a Ca^{2+}-dependent manner (Caccamo et al., 2004; Pajak and Orzechowski, 2007a). Moreover, CLU is involved in tumorigenesis and regulation of cell life and death (Ammar and Closset, 2008; Caccamo et al., 2003; Chayka et al., 2009; Leskov et al., 2003; Liu et al., 2009; Redondo et al., 2007; Rizzi et al., 2009; Scaltriti et al., 2004; Yang et al., 2000; Zhang et al., 2005). Thus, the answer to the frequently asked question of how to explain the idiosyncrasy of CLU, may find an explanation studying Ca^{2+} and its role on the normal and transformed cells. Since Ca^{2+} is also indicated to play significant role in protein degradation, the CLU deprivation has a special merit in well proven pro- and antiapoptotic effects. It is worth to mention that intracellular Ca^{2+} fluxes are tightly controlled both in time and space since a number of catalytically active proteins critical for cell fate depend on Ca^{2+}. Additionally, adhesiveness and other CLU activities seem to be associated with Ca^{2+}, but these attributes need further exploration. The position of Ca^{2+} in control of CLU activities is principally dependent on Ca^{2+} channels and their regulation. In this chapter, we took on the task to give an explanatory note to the well-known features of CLU with respect to Ca^{2+} waves. The issue of Ca^{2+}-dependent regulations of CLU activity was the main objective in our effort to provide the reader with up to date information, pros and cons, as well as the gaps that remain to be perceptive in Ca^{2+}–CLU relations.

II. CA^{2+} SIGNAL

In contrast to extracellular milieu (1 mM), intracellular cytoplasmic Ca^{2+} concentration is kept low ($\sim$100 nM) while mitochondria and endoplasmic reticulum (ER) pile up Ca^{2+} (millimolar range) from where it is occasionally released upon stimuli mediated by second messengers (i.e., inositol triphosphate, IP3). Besides, Ca^{2+} is known to promote its own release through the ryanodine Ca^{2+} channels. Cell membrane is impermeable to ionic calcium except for plasma membrane calcium channels (PMCC) which are differently gated (by voltage or ligands). Additionally, Ca^{2+} is forced out from the cell by plasma membrane pumps (Ca^{2+}-ATPases) or forced out back to ER by sarcoplasmic reticulum Ca^{2+}-ATPases (SERCA). Specific calcium ion channel blockers were used to study the activity of respective channels. Ca^{2+} enters cytoplasm transiently (for seconds or minutes), as there is a favorable concentration gradient. Instantaneously, Ca^{2+} triggers a variety of signals, including activation/inactivation of numerous regulatory and catalytically active proteins (enzymes, transcription factors, chaperones, etc.). Importantly, CLU/Apo J belongs to the group of, in some way, Ca^{2+}-modulated proteins. Although we know much about Ca^{2+} fluxes, it is indicative that the definitive criteria which determine cell fate in response to Ca^{2+} are not known accurately. Possibly, Ca^{2+} concentration may vary in time, even may oscillate (Ca^{2+} influx from intracellular stores and subsequent extracellular Ca^{2+} entry) offering a room for somewhat diverse outcomes. Frequently, Ca^{2+} fluxes bring opposite consequences, such as cell proliferation versus differentiation, viability versus death, apoptosis versus necrosis. Probably, partitioning of Ca^{2+} entry into separate cellular compartments plays a significant role in the process. In any case, Ca^{2+} signaling is often examined when apoptosis is the objective. It has been shown that discrete Ca^{2+} signals trigger either the intrinsic, caspase-dependent apoptosis, or caspase-independent, calpain-mediated apoptosis (Pinton *et al.*, 2008; Tagliarino *et al.*, 2001). Overall, these observations suggest that calcium homeostasis is vital for cell survival, including tumor cells. The details how Ca^{2+} signals affect particular type of tumor are critical if one wish to defeat immune escape and antiapoptosis of cancer cells. Links between the Ca^{2+} signal, CLU activity, and chemoresistance designate the approach that should be cautiously inspected.

III. CA^{2+} CONTRIBUTION TO CLU-MEDIATED EFFECTS

Presumably, high levels of calcium in serum may promote the growth of potentially fatal cancers (Liao *et al.*, 2006). Among variety of cancers, the correlation between calcium ion (Ca^{2+}) and cancer incidence in man was observed for prostate cancer (Skinner and Schwartz, 2009). Actually, this suggests that prospective association between serum calcium and prostate cancer mortality exist. The G-protein-coupled receptors (GPCR) activated by Ca^{2+} were identified in prostate cancer cells (Lin *et al.*, 1998). Alongside these sensors, prostate cancer proliferation is under control via Ca^{2+} entry through the Ca^{2+}-activated K^+ channels (Lallet-Daher *et al.*, 2009).

In contrast, in many other cell types (including cancers), it is thought that high intracellular Ca^{2+} makes cancer cells primed to cell death (Tagliarino *et al.*, 2001, 2003). Surprisingly, CLU translocates from cytosol to nucleus at low intracellular Ca^{2+} levels which correlates with reduced cell viability and anoikis in prostate or classical apoptosis in colorectal cancer cells (Caccamo *et al.*, 2004; Pajak and Orzechowski, 2007a). It was exemplified by extra-cellular calcium deprivation that leads to reduced viability of astrocytes by apoptosis (Chiesa *et al.*, 1998). Besides, since secretory activity relies on Ca^{2+} influx (Catterall, 2000), intracellular CLU accumulation might be brought about by low intracellular calcium. Under certain stress conditions, CLU may evade the secretion pathway and accumulate in cytosol (Nizard *et al.*, 2007). Thus, the alternative route for CLU location has been postulated in parallel to exon skip or alternate splicing of mRNA.

It was not a simple coincidence that CLU presence was first demonstrated in seminiferous tubules. Findings of D'Agostino group clearly showed that activity of Sertoli cells which control spermatogenesis in testis is affected by intracellular Ca^{2+} and that selection of spermatocytes or spermatogonia to methoxyacetic acid-induced apoptosis is associated with high intracellular CLU (Barone *et al.*, 2003, 2005). As was mentioned, several voltage-gated calcium channels (VGCC) have been identified by use of specific calcium ion channel blockers (N-, L,- and P/Q-type) in Sertoli cells in *in vitro* and *in vivo* studies. The P/Q-type VGCC is unique for its cellular presence and physio-logical effects. By comparison, L-type VGCC modulate a laminin-dependent Ca^{2+} influx (Taranta *et al.*, 2000), while N- and P/Q-type VGCC are involved in the regulation of protein secretion (Fragale *et al.* 2000; Taranta *et al.*, 1997). P/Q-type VGCC channels are located at plasma membrane of Sertoli cells lining the basal lamina where blood–testis barrier is set up. Interestingly, male gametes linked to Sertoli cells do not express this type of Ca^{2+} channel. The experimental evidence showed that CLU protein was secreted by Sertoli cells in response to P/Q VGCC channel activation. Subse-quently, elevated CLU levels were detected in gametes undergoing pro-grammed cell death (Barone *et al.*, 2003, 2005). The detailed molecular mechanism of this reaction is not known at present, although it is admitted that CLU from Sertoli cells somehow entered male gametes and facilitated apoptosis. Actually, intact junctions between Sertoli cells and spermatocytes were essential since CLU secreted was unable to penetrate intact germ cells. Although spontaneous apoptosis effects up to three-fourth of potential male gametes, even more is deleted in the genotoxic conditions. This process limits the clonal expansion of germ cells to the number that Sertoli cells are able to support. It probably prevents the spread of aberrant spermatozoa too.

Recently, it was shown that L-type VGCC allow heavy metal ions Zn^{2+} and Cd^{2+} to enter germ cells (Kaisman-Elbaz *et al.*, 2009). Once permeated, they trigger apoptosis. It comes across with high levels of Zn^{2+} and Cd^{2+}

found in testis. Notably, Zn^{2+} and Cd^{2+} cell influx in depolarizing conditions enhanced CLU expression and secretion, followed by cell death. Conversely, attenuation of CLU or Zn^{2+} and Cd^{2+} intracellular entry reversed eradication of germ cells. This study reminds us that EDTA is a nonspecific Ca^{2+} chelator ridding of Zn^{2+} and Cd^{2+} prior to Ca^{2+} removal and pointing to cautious interpretation of EDTA-based experiments (Kay, 2004). Much remains to be learned on the role CLU in tumorigenesis of testis.

IV. CA^{2+} INFLUX THROUGH TRPM-2 CHANNELS AND CLU LOCATION

Initially, one of the names given to CLU was TRPM-2 (testosterone repressed prostate message 2), abbreviation identical to channel enzyme TRPM-2 (transient receptor potential melastatin 2). Ten years have passed from discovery of TRPM family of cation channels which are gated by Ca^{2+}, ADP-ribose (ADPR), H_2O_2, and NAD^+ (Nagamine *et al.*, 1998). They were identified in CLU-rich tissues such as brain (Nagamine *et al.*, 1998) and prostate (Wang *et al.*, 2007). ADPR coalesces with the intracellular C-terminal Nudix-box, a homology domain to enzymes with pyrophosphatase activity. It was demonstrated that Ca^{2+} and ADPR might together regulate intracellular Ca^{2+} influx, and Ca^{2+} facilitates gating of TRPM-2 channel by ADPR. The molecular mechanisms of Ca^{2+}-, H_2O_2-, or NAD^+-dependent TRPM-2 activation as well as signaling pathways involving TRPM-2 activation are not fully understood. It is evident from several studies that TRPM-2 activation leads to induction of cell death but its intracellular effectors were not revealed, either. TRPM family of cation channels are voltage operated, although they do not possess voltage sensor (Voets *et al.*, 2004). The common feature of TRPM channels is N-terminal TRPM-homology domain, indispensable for gathering and anchorage of protein channel complex (Perraud *et al.*, 2003). Three major subfamilies are canonical (TRPC), melastatin-related (TRPM) and vanilloid-related (TRPV) channels, although remote subfamilies of polycystin (TRPP), mucolipin (TRPML), ankyrin (TRPA), and no mechanoreceptor potential C or NOMPC (TRPN)-related channels were also identified. The expression of TRPM and TRPV channels was found in prostate continuum (Wang *et al.*, 2007). Of all TRPM subtypes, TRPM8 expression was the highest while for the TRPV subfamily, TRPV4 was most plentiful in rat prostatic tissue. The function of above-mentioned TRPM and TRPV channel members is uncertain in prostatic tissue. Anyway, it was frequently reported that intracellular CLU location is tightly regulated by Ca^{2+}. In contrast to typical VGCC which are selective for Ca^{2+} the TRPM and TRPV subtypes are nonspecific cation channels. It suggests that discrepancies observed in CLU studies when different Ca^{2+} chelators were used could be associated with different chelator affinity to cations. For example, EDTA has a far higher affinity for transition metals than it does for calcium, so it chelates transition metals with the displacement of calcium. EGTA specificity favors Ca^{2+} instead of transition metals. In our study where EDTA and EGTA were exploited in equimollar concentrations, merely EDTA could elicit anoikis in human COLO 205 adenocarcinoma cell line. Cytotoxicity of EDTA was accompanied by CLU nuclear setting (Pajak and Orzechowski, 2007a). Previously, Caccamo et al. (2004) showed identical reaction in human prostate cells when BAPTA AM was utilized to eliminate intracellular calcium. BAPTA AM is a well-known intracellular calcium blocker that enters the cell, is cleaved and trapped irreversibly as charged Ca^{2+} chelator. Even then, at high concentrations BAPTA can trigger apoptosis (Caccamo et al., 2005; Pu and Chang, 2001) and increased CLU promoter activities (Araki et al., 2005). It is not clear, if intracellular Ca^{2+} depletion, or Ca^{2+} influx from the outside elicit cell death. The risk of BAPTA toxicity at higher doses may also be attributed to Ca^{2+} deprivation from ER as BAPTA can enter this compartment. Altogether, these results demonstrate two faces of Ca^{2+} and Ca^{2+}-associated control of CLU function. Secreted form of CLU (sCLU) stimulated by Ca^{2+} sustains viability upon cell stress, thus playing a role of chaperone, whereas its intracellular form (nCLU) is induced by drop in Ca^{2+} and favors cell deletion. Opposite effects observed from chelation studies suggest that Ca^{2+} concentration substantially affects cell proliferation. For this reason, CLU contribution to mitogenesis is a subject matter of extensive research.

V. CA^{2+} AND CELL PROLIFERATION

In the past, several reports documented Ca^{2+} involvement in mitogenesis (Mailland et al., 1997). Boynton et al. (1974) described how changes in extracellular Ca^{2+} concentration impinge on proliferation of 3T3 fibroblasts in serum supplemented medium. A minimum 0.05 mM of extracellular Ca^{2+} was indispensable to support cell divisions while 0.5 mM was optimal. Similar results were obtained by other authors when serum mitogens (growth factors) were examined (insulin, Monaco et al., 2009), insulin-like growth factors (IGFs, Kojima et al., 1988), fibronectin (Illario et al., 2008), platelet-derived growth factor (PDGF, Ma et al., 1996), epidermal growth factor (EGF, Peppelenbosch et al., 1992), and fibroblast growth factors (FGFs, Munaron et al., 1995). They act on target cells by binding to membrane receptors associated with kinase activity (receptor tyrosine kinase, RTK) while other peptides (i.e., bradykinin) act through GPCR. By assembly with cognate receptors, mitogens elicit cascades of intracellular signals. Mitogen-activated protein kinases (MAPK, ERK1/2, Illario *et al.*, 2008) and phosphatidylinositol 3 kinase/Akt (Khundmiri *et al.*, 2006) play decisive roles by starting replicative processes. Early signals lasting from seconds to tens of minutes are apparently of paramount importance for ionic events. They include activation of transporters and channels as well as ionic redistribution between subcellular compartments (Barbiero *et al.*, 1995). By Ca^{2+}-induced Ca^{2+} release from internal stores (ryanodine receptors) or Ca^{2+} permeation from extracellular medium the sustained increase in Ca^{2+} might be achieved. Consequently, capacitative and noncapacitative Ca^{2+} entry for at least 1 hour allows for the progression from G_0 to G_1 of cell cycle (Estacion and Mordan, 1993; Fig. 1).

Recently, cytofluorometrical and electrophysiological evidence for Ca^{2+} influx showed the substantial role of ion channels in cell-cycle regulation, including L-type VGCC (Monaco *et al.*, 2009). Transient receptor potential (TRPC-1 and TRPM-7) channels are also indicated to mediate Ca^{2+}-dependent stimulation of cell proliferation (Abed and Moreau, 2009; El Hiani *et al.*, 2009). Patch-clamp electrophysiological recordings (whole cell mode, single channel) were widely used to extend our grasp on the pattern of cytosolic calcium. Calcium channel blockers (nifedipine, verapamil, SKF96365A), chelators (EDTA, EGTA, BAPTA AM), ionophores (calcimycin/A23187), and Ca^{2+}-sensitive dyes (fura 2) were employed as tools. Serious limitation of these techniques is the duration of experiment (less than 1 h). Even then, calcium response to mitogenic stimuli is biphasic, with initial rise (transient) followed by a long-lasting lower phase featured by several oscillations. It is believed, that initial burst originates from release of Ca^{2+} from intracellular stores since it occurs in the absence of extracellular Ca^{2+}. It is not the case for subsequent plateau where external source of Ca^{2+} is essential.

VI. IS CLU A CA^{2+}-DEPENDENT PRO- OR ANTIAPOPTOTIC GENE?

Overexpression of CLU was either observed to suppress apoptosis/ameliorate viability in some tumors (Ammar and Closset, 2008; Liu *et al.*, 2009; Redondo *et al.*, 2007; Zhang *et al.*, 2005) or to be proapoptotic/to depress survival in other multiple tumor cell lines (Leskov *et al.*, 2003; Yang *et al.*, 2000). Actually, CLU is a short-lived protein, with the average turnover of less than 2 h (Rizzi *et al.*, 2009). Opportunely, this half-life corresponds to the time window where one can keep an eye on Ca^{2+} intracellular fluxes.

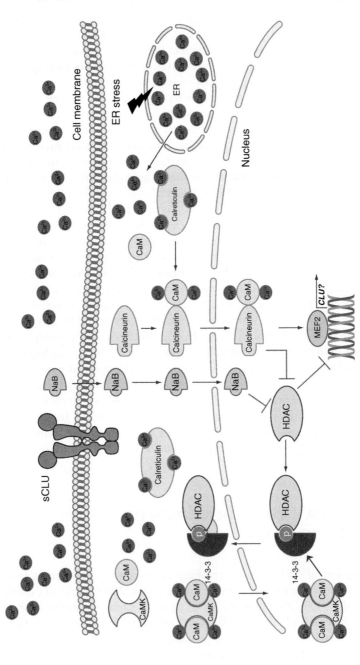

Fig. 1 Suggested role of Ca^{2+} ions in the control of cell proliferation. Serum mitogens acting through membrane receptors (RTK, GPCR) activate transporters and channels as well as ionic redistribution between subcellular compartments. Mitogen-activated protein kinases (MAPK, ERK1/2) and phosphatidylinositol 3 kinase/Akt are stimulated to target protooncogenes. By Ca^{2+}-induced Ca^{2+} release from internal stores (ryanodine receptors) or Ca^{2+} permeation from extracellular medium the sustained increase in Ca^{2+} is achieved. Consequently, the progression from G_0 to G_1 of cell cycle might proceed since nCLU stays dormant (pnCLU) in the cytoplasm. (See Page 2 in Color Section at the back of the book.)

Irrespective to its short half-life, nCLU was often reported to act as an antioncogene protein and *CLU* is consequently summoned up as the tumor suppressor gene (Caccamo *et al.*, 2003; Chayka *et al.*, 2009; Scaltriti *et al.*, 2004). Furthermore, CLU-dependent cell-cycle arrest directly correlates with the drop of intracellular Ca^{2+} and its transfer to nucleus (Caccamo *et al.*, 2004; Pajak and Orzechowski, 2007a). CLU behavior bears the resemblance of transcription factors who once activated translocate and regulate gene activity (Batsi *et al.*, 2008). It is predominantly noticeable in *p53* "loss of function" cancer cell lines (Batsi *et al.*, 2008; Chen *et al.*, 2004; Mallory *et al.*, 2005; Markopoulou *et al.*, 2009) probably because *CLU* gene is negatively regulated by wild-type p53 (Criswell *et al.*, 2003). The latter possibility is still a matter of debate, although Yang *et al.* (2000) demonstrated that CLU interacts with DNA repair machinery in irradiated cells. At this point, CLU induced cell-cycle arrest and triggered the programmed cell death. A number of papers has been given testimony of nCLU-mediated proapoptotic activity (Caccamo *et al.*, 2003; Leskov *et al.*, 2003; Yang *et al.*, 2000) while some emphasized the causal relationship between cell-cycle withdrawal and apoptotic cell death (Chayka *et al.*, 2009; Markopoulou *et al.*, 2009; Scaltriti *et al.*, 2004; Shannan *et al.*, 2006a,b; Fig. 2). Taken together, CLU is implicated in various cell functions, including DNA repair, cell-cycle regulation, and apoptotic cell death. Importantly, above-mentioned activities are differently controlled by cytosolic Ca^{2+}. It signifies the existence of cell death promoting pathway which is an attractive alternative to caspase-dependent cell deletion. Possibly, robust cell death is elicited by calpain proteolytic system (Orzechowski *et al.*, 2004). Micro- or millimolar Ca^{2+} concentrations trigger μ- and m-calpain, respectively. The system is tightly controlled by calpastatin, however, high cytosolic Ca^{2+} overcomes calpastatin inhibition (Orzechowski *et al.*, 2003). As other proteins, CLU is obviously a target of transcriptional and posttranslational regulation. In this process, however, not Ca^{2+} but proteasome and lysosome systems are indicated in CLU proteolytic cleavage. Actually, it is widely accepted that "old" CLU is polyubiquitinated and degraded by proteasome 26S (Balantinou *et al.*, 2009; Rizzi *et al.*, 2009). Nevertheless, a great deal of data comes from studies where proteasome inhibitor MG132 was used to promote CLU accumulation and to activate *CLU* gene expression (Balantinou *et al.*, 2009; Rizzi *et al.*, 2009). It has to be stressed, however, that MG132 at concentrations used almost entirely blocks calpain activity as well (Tsubuki *et al.*, 1996). Previously, CLU was described as the potent NF-kappaB (NF-κB) repressor that by unknown mechanism leads to elevated stability of IkappaB (IκBα), innate NF-κB inhibitor (Santilli *et al.*, 2003). However, when breast cancer cells were treated with chemotherapeutic drug doxorubicin (Doxo) combined with histone deacetylase inhibitor (HDACI) sodium butyrate (NaB) both stabilized CLU but at the same time a second proteasome target,

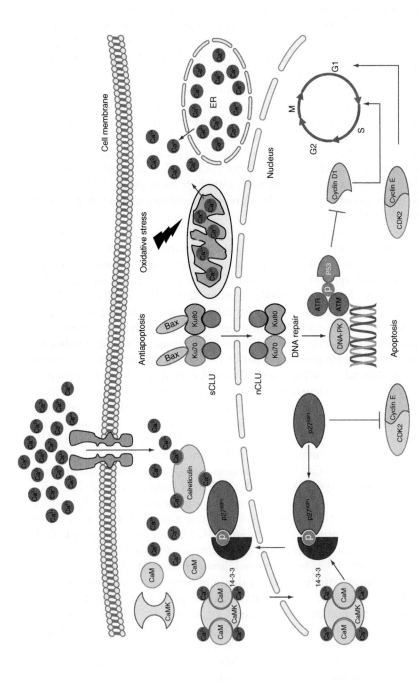

Fig. 2 Suggested role of Ca^{2+} ions in the control of apoptosis. Oxidative stress leads to modifications of calcium ion channel proteins, leakage of Ca^{2+} from intracellular stores (mitochondria, ER), and from the extracellular space. At the same time, DNA lesions are generated. It is hypothesized that

IκBα was degraded (Ranney *et al.*, 2007). This apparent inconsistency is explicable if one admits that CLU was inhibited by calpain proteolytic pathway (Fig. 3). In such circumstances, cytosolic Ca^{2+} fluctuations may considerably affect CLU levels. Rise in Ca^{2+} would mean less intracellular CLU (calpain-dependent degradation) whereas fall in Ca^{2+} would signify the elevated CLU level. Additionally, CLU level rises in cellular stress, including oxidative insult (Araki *et al.*, 2005; Stokes *et al.*, 2002). Although not completely understood, reactive oxygen species (ROS) release Ca^{2+} from ER stores/extracellular milieu pointing to stress-dependence of cytosolic calcium concentration (Orzechowski *et al.*, 2003). This issue should be urgently addressed since CLU degradation by Ca^{2+}-controlled calpain in cancer cells would overcome its antiproliferative and antitumor effects. HDACI-s attracted special attention owing to their antitumor activity and superior differentiation induction. Butyrate, a short-chain fatty acid (SCFA) released during microbial digestion of fiber in the bowel facilitated extrinsic apoptosis (Pajak and Orzechowski, 2007b; Pajak *et al.*, 2009). As other observed, this event occurs despite the concurrent CLU overexpression (Liu *et al.*, 2009). Anyway, the proapoptotic activity of HDACI-s overrides hampering effect of sCLU on intrinsic apoptosis (Zhang *et al.*, 2005) and peaks at extensive cell death due to synergy with tumor inhibitors (Ranney *et al.*, 2007). Aforementioned observations were corroborated by *in vitro* and *in vivo* studies with antisense oligonucleotides (ASO) or siRNA targeting *CLU* gene. Data showed that elimination of full-length CLU mRNA further sensitized tumor cells to chemotherapeutic drugs (July *et al.*, 2004) or to HDACI (Liu *et al.*, 2009). Even then, the application of ASO such as OGX011 when used alone could not delete prostate and breast cancer cells (Chen *et al.*, 2004).

VII. CA^{2+} HOMEOSTASIS AND CLU TRANSCRIPTION

The details of control of regulation of *CLU* gene transcription call for reminding the distinctive origin of secretory (sCLU) and nuclear (nCLU) clusterin. If CLU mRNA is truncated (alternative splicing) the translation product is shorter (49 kDa) as it lacks the leader sequence (LS). It remains inactive in the cytoplasm, unless the NLS is "put into sight." Subsequently, it

---

depending on Ca^{2+} concentration either sCLU binds Ku70–Bax (at elevated Ca^{2+} concentration) to maintain cell viability or nCLU binds Ku70 (at low Ca^{2+} concentration) to compromise Bax–Ku70 interactions. Former case leads to antiapoptosis, whereas the latter evokes apoptosis. Cell-cycle checkpoint G1/S is extended for DNA repair unless the ATM/ATR kinases stimulate p53 protein to generate stress signals. (See Page 3 in Color Section at the back of the book.)

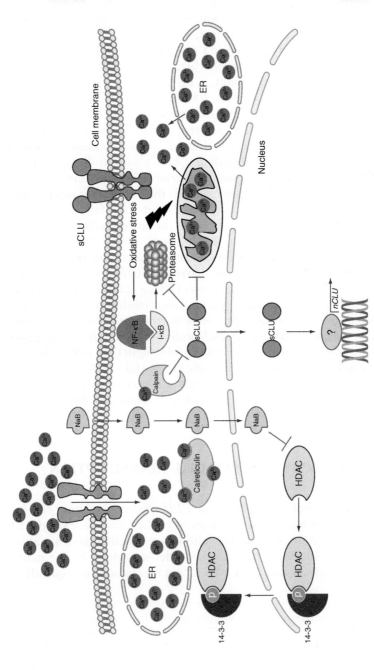

Fig. 3 Suggested role of Ca^{2+} ions in the control of cell survival. CLU inhibits IκB degradation by ubiquitine–proteasome system, although at the same time CLU might be the subject for proteolysis. It is conceived, that the alternative Ca^{2+}-dependent calpain system hits CLU, so NF-κB cannot be repressed any longer. NF-κB is a transcription factor of antiapoptosis, so once activated it stimulates activity of several genes coding antiapoptotic proteins. This picture exemplifies the belief that elevated Ca^{2+} represses nCLU (it remain in dormancy) which otherwise acts as antioncogene. (See Page 4 in Color Section at the back of the book.)

is supposed to be posttranslationally altered to 55 kDa cell death protein (nCLU), although several reports do not confirm this additional CLU processing (Scaltriti *et al.*, 2004). Functional analysis of nCLU showed that ultimately C-terminus evokes apoptosis (Leskov *et al.*, 2003). It remains uncertain whether increase or decrease of Ca^{2+} is crucial; nonetheless, the turmoil in Ca^{2+} homeostasis leads to change in *CLU* gene activity. CLU is induced in cancer and normal cells by ionizing radiation (IR), hypoxia/hyperoxia, stress-inducible cytokines, and other less defined factors which accompany profound trauma (Araki *et al.*, 2005; Han *et al.*, 2001; Trougakos and Gonos, 2004; Zellweger *et al.*, 2002). Nowadays, the chase started to find common path that would elucidate why cells liberate sCLU in order to survive and why do they die if nCLU is upregulated. Notably, intracellular sCLU accumulation was also observed when this protein evaded the secretion pathway (Nizard *et al.*, 2007). Araki *et al.* (2005) hypothesized that production of ROS guides the Ca^{2+} waves. ROS causes lipid peroxidation (plasma membrane) and protein modifications with damage of Ca^{2+} channels and Ca^{2+} transporters. Free Ca^{2+} mounts up to the point where either *CLU* gene is activated and sCLU is secreted or Ca^{2+} becomes poisonous to start cell death by apoptosis. In the first case, sCLU is released and acts as a cell debris scavenger (Tschopp *et al.*, 1993). Thus, by ridding of the mediators of inflammatory reaction sCLU stops further production and secretion of proinflammatory cytokines which are known to promote procoagulant phenotype of endothelium (disseminated intravascular coagulopathy, DIC)—basis for multiple organ failure (MOF), a fatal syndrome associated with terminal stages of cancer disease. The issue of Ca^{2+}-dependent control of transcription has important role in regulating programmed cell death (Mao *et al.*, 1999; Youn *et al.*, 1999). The mechanisms of control were first described in muscle cells where conversion from fast to slow muscle fibers is mediated by calcineurin (Ca^{2+}/calmodulin-dependent phosphatase) (Chin *et al.*, 1998). It involves the Ca^{2+}-dependent dissociation of class II histone deacetylases (HDAC-s) from DNA binding sites for myocyte enhancer binding factor 2 (MEF2) (Lu *et al.*, 2000). Once activated, MEF2 transcription factor amplifies genes essential for myogenesis (McKinsey *et al.*, 2002). Obviously, Ca^{2+} also activates calmodulin-dependent kinases (CaMK) which phosphorylate free HDAC-s for the export from the nucleus with 14-3-3 proteins. Taken together, the effects of Ca^{2+} are similar to those brought about by HDACI-s, which have been already shown to promote CLU expression (Fig. 4). Whether Ca^{2+}-dependent MEF2 is involved in the regulation of *CLU* gene transcription is not known at present, even so looks like it is a very attractive goal for exploration. Moreover, it has been shown that *CLU* gene is silenced by methylation in prostate cancer cells (Rauhala *et al.*, 2008). This report corroborates the results of Rizzi *et al.* (2009), having shown that nCLU controls fate of

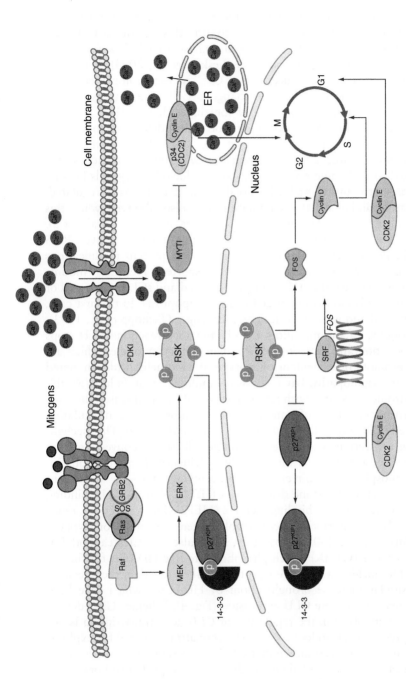

Fig. 4 Suggested role of Ca^{2+} ions in the control of *CLU* gene expression. Endoplasmic reticulum stress provides Ca^{2+} ions with calreticulin as vehicle to target calmodulin (CaM). Consequently, calmodulin-dependent proteins as calcineurin phosphatase and calmodulin-dependent kinase (CaMK) are activated. Calcineurin allows dissociation of class II histone deacetylases (HDAC) from DNA, thus providing access for transcription factors (i.e., MEF2). At the same time, CMK phosphorylates HDAC to set it aside from nucleus with 14-3-3 protein. Similar effect is obtained with sodium butyrate (NaB) which is a strong HDAC inhibitor. (See Page 5 in Color Section at the back of the book.)

prostate cancer cells. Transient overexpression of CLU slowed down cell-cycle progression and retarded growth of cancer rather than normal prostate epithelial cells (Bettuzzi *et al.*, 2002). For this reason, nCLU is considered as a putative tumor suppressor gene product that might play pivotal role in the inhibition of oncogenesis apart from *CLU* gene loss of function. Overall, these observations shed more light on the value of epigenetically regulated genes and specify some role of Ca^{2+} in the acetylation and/or methylation of *CLU* gene that localizes on chromosome 8 (8p21) of human genome (Fink *et al.*, 1993; Purrello *et al.*, 1991), a portion of the genome which is frequently deleted in prostate cancers.

VIII. CLU TRANSLOCATIONS IN RESPONSE TO CA^{2+} SIGNAL

Intracellular shifts of CLU between the cytoplasm and nucleus and vice versa, as well as dissimilar expressions of nCLU in aforementioned compartments were demonstrated to be important for cellular decision to live or let die (Araki *et al.*, 2005; Caccamo *et al.*, 2003, 2005; Miyake *et al.*, 2003; Nizard *et al.*, 2007; Pajak and Orzechowski, 2007a). CLU translocations seem to be critically important for biological effects while Ca^{2+} is an apparent trigger. Frequently, CLU is induced in response to genotoxic or ER stress, but alternative transcription initiation or alternative splicing are not needed (Nizard *et al.*, 2007; Rizzi *et al.*, 2009). The overall picture of CLU functions is therefore multifaceted and requires careful dissociation between different stimuli, CLU forms, and Ca^{2+} resources. Interesting data were presented by authors who make use of calcium moderators. Araki *et al.* (2005) found that BAPTA AM (intracellular Ca^{2+} chelator), thapsigargin (inhibitor of SERCA), and β-lapachone (increases cytoplasmic Ca^{2+} from ER stores) when used one after the other in different sequences led to rather different results. Authors concluded that moderate rise in Ca^{2+} upon low doses of traumatic agent (IR) activated *CLU* gene and sCLU formation to protect cells from insult, while high doses of IR stimulated *CLU* gene with nCLU death protein as a major product. Alterations in intracellular Ca^{2+} markedly induce sCLU, seemingly to avoid outcomes of cell stress. Administration of BAPTA AM did not prevent thapsigargin-induced lethality, but it did protect against β-lapachone cytotoxicity (Tagliarino *et al.*, 2001). Astonishingly, in high doses, BAPTA AM also became cytotoxic and enable to induce *CLU* gene. The role of BAPTA AM in the latter response is unclear, but Caccamo *et al.* (2005) revealed that PNT1A cells kept in Ca^{2+} supplemented medium were sensitized by BAPTA AM to nCLU-induced apoptosis. On the other hand, at lower doses, Tagliarino *et al.* (2003) and

Araki *et al.* (2005) both observed preventive effect of BAPTA AM on β-lapachone-mediated apoptosis in MCF-7/WS8 cells. In addition, they unraveled that β-lapachone treatment activated μ-calpain rather than caspases and executed programmed cell death with PARP cleavage to a unique 60 kDa fragment. Interestingly, when MCF-7/WS8 breast cancer cells were studied by Araki *et al.* (2005), it became apparent that β-lapachone activates NAD(P)H:quinone oxidoreductase-1—NQO1 often upregulated in tumor cells (Marin *et al.*, 1997). Oxidative stress and futile β-lapachone quinon: hydroquinon cycle driven by NQO1 depletes NADH, NADP(H), and ATP. As a result, mitochondrial membrane potential collapses, Ca^{2+} escapes mitochondrial matrix, and Ca^{2+} signal arrives to cytoplasm. By what mechanism does BAPTA AM protect, and by what it facilitates apoptosis was ambiguous until different Ca^{2+} membrane-impermeant inhibitors (membrane-impermeant BAPTA and heparin) were injected by electroporation into cytosol (Pu and Chang, 2001). Contradictory effects (higher BAPTA concentration induces while lower inhibit apoptosis) seems to be due to its dual effect on subcellular Ca^{2+} distribution. Besides suppressing the Ca^{2+} rise in cytosol, BAPTA can also enter into the ER to reduce the free Ca^{2+} level within. The depletion of Ca^{2+} in ER may stimulate apoptosis and thus would defeat the protection effect of BAPTA AM in suppressing the cytosolic Ca^{2+} increase. CLU is supposed to be a tool of cell deletion.

IX. ER STRESS, UNFOLDED PROTEIN RESPONSE, CA^{2+}, AND CLU

Heat, reducing, or oxidative insult all can cause severe disorder in protein maturation (Rose and Doms, 1988). Ca^{2+} depletion from ER also disrupts protein folding leading to unfolded protein response (UPR) (Rowling and Freedman, 1993). A series of signal transduction cascades is elicited, especially in professional secretory cells where high level of secretory protein synthesis is required (breast, prostate). Ca^{2+} occurrence in ER is remarkably high (5–10 mM), as Ca^{2+} is vital for control of calnexin (Cnx) and calreticulin (Crt) activities. These chaperones possess lectin-binding site that specifically select proteins for secretion or posttranslational modifications in Golgi complex. Proteins that exit ER are correctly folded, while misfolded are hold on for proteolytic disposal. It is indispensable that quality-control criteria have to be met prior to export of any protein from ER. Cnx (90 kDa) is ER membrane bound, whereas Crt (60 kDa) resides in the ER lumen. Both proteins associate with most, if not all, glycoproteins that pass through the ER (Williams, 2005). Cnx/Crt protein complex lasts until the protein folding is successfully completed. ATP and Ca^{2+} are two cofactors engaged in substrate binding for calnexin. In turn, calreticulin has a low affinity but it buffers Ca^{2+} efficiently, so Crt is a major Ca^{2+}-binding (storage) chaperone in ER (Michalak et al., 2009). Previously we mentioned, that CLU level rises in ER stress and that sCLU is ubiquitously expressed and set up for secretion. Once activated by stress, cells push out sCLU so very low levels of CLU remain inside the cell, but if secretion is stopped (low Ca^{2+}), the cell inevitably will die. A bulk of evidence suggests that the most likely candidate to be misfolded at some point in ER stress is sCLU. Two polypeptide chains of sCLU have to be linked by five disulfide bonds in more oxidizing environment than cytosol. The formation of disulfide bonds is catalyzed in ER by protein disulfide isomerase (PDI) while the reaction is controlled by Cnx/Crt upon Ca^{2+} abundance. Elegant study performed by Losh and Koch-Brandt (1995) brought the evidence how psCLU is processed in ER and that reducing microenvironment is able to retain this protein in ER unless more oxidized lumenal redox state is reestablished. sCLU is a soluble marker for apical exocytosis and ER stress does not affect secretion of properly folded and fully glycosylated form of sCLU (Losh and Koch-Brandt, 1995). We could not find any available data on the position of Ca^{2+}-dependent Cnx/Crt effect on CLU activity. However, it is admitted that stress-induced retrotranslocation of CLU from ER to cytosol is associated with polyubiquitination similarly to other misfolded proteins (Nizard et al., 2007). The contrasting phenotypes of Cxn and Crt deficiency in mice (Denzel et al., 2002; Mesaeli et al., 1999) indicate that despite structural similarities these chaperones do not have common functional characteristics and probably play different roles in development. Furthermore, mice phenotypes are more severe than those of cultured cells ($Crt^{-/-}$ knockout mice die in embryonic life, $Cnx^{-/-}$ are viable but have reduced survival), whereas cultured cells endure the absence of Cnx/Crt (Molinari et al., 2004). With regard to CLU, it should be emphasized that overexpression of activated Crt rescues $Crt^{-/-}$ mice (Guo et al., 2002). Given that activation of calcineurin is stimulated by Ca^{2+} release from ER and that ER Ca^{2+} is maintained by Crt it seems likely that Crt is an upstream regulator of calcineurin in Ca^{2+} signaling (Lynch and Michalak, 2003). Although it is highly speculative, the connections between epigenetically regulated *CLU* gene, calcineurin, and calreticulin are feasible and it is tempting to examine if any relationships exist between Ca^{2+}, Cnx/Crt, and CLU. Equally important question to ER stress is related to the redox disorder as a possible causal factor of Ca^{2+}-mediated and CLU-associated apoptosis (Markopoulou et al., 2009).

As described above, associations between Ca^{2+}/Ca^{2+}-controlled proteins, cancer cells and expression of *CLU* gene are indicative for CLU isoforms as modulators of cell fate. There exist a limited number of studies performed with calcium chelators, calcium ionophores, and Ca^{2+} ion channel blockers that could help to establish the definite role of Ca^{2+} in CLU activities.

Certainly, Ca^{2+} signal is vital for the control of a cellular constituent, and some of them are known to affect CLU expression and localization. Both extracellular and intracellular Ca^{2+} depletion as well as excess of Ca^{2+} impaired viability of cancer cells suggesting that Ca^{2+} concentration should remain within narrow physiologically acceptable limits. Any long-term disturbances in Ca^{2+} homeostasis seem to affect cell-cycle control and sCLU and nCLU are most likely proteins involved in this regulation.

X. CA^{2+}, CLUSTERIN, AND DNA REPAIR

Opposing results obtained from studies on tumor cell lines with regard to pro- and antiapoptotic activity of CLU imply that these effects might be regulated by Ca^{2+} fluxes. Intracellular Ca^{2+} abundance drives CLU secretion, whereas lack of Ca^{2+} leads to CLU accumulation, nuclear translocation, and cell death (Scaltriti et al., 2004). CLU might become cytotoxic when accumulates in high amounts intracellularly either by direct synthesis or by uptake from the extracellular space (Trougakos and Gonos, 2004). In turn, secreted CLU protects cells from the hostile microenvironment and reduces risk of genotoxicity. Accordingly, cells differ in their strength to secrete sCLU and those less efficient could be less protected from genotoxic insult. sCLU suppressed p53-dependent stress signaling by maintaining Ku70–Bax protein complex in the cytoplasm (Trougakos et al., 2009). It seems likely, that by similar mechanism the proinflammatory cytokine IL-6 promotes immune escape of human colon cancers by stabilizing the Bax–Ku70–sCLU complex (Pucci et al., 2009). Conversely, nCLU was identified as the first stress-inducible protein that binds Ku70, forming a trimeric protein complex with Ku80, prior to assembly with DNA-dependent protein kinase (DNA-PK) activated by DNA strand breaks (Fig. 2). It is believed that sCLU suppresses apoptosis by competing for Ku70 with Bax protein (Shannan et al., 2006a). In contrast, the lack of LS permits nCLU to promote apoptosis (Leskov et al., 2003). To decipher Ca^{2+} position in CLU contribution to DNA repair additional studies are needed, even though at least spontaneous DNA repair is calcium dependent (Ori et al., 2005). Given that, Ca^{2+} message heralds intrinsic apoptosis it is confusing that Ca^{2+} depletion is associated with nCLU activation, nuclear location, and cell death. However, Ca^{2+} is not essential for apoptosis, with the exception of DNA oligonucleosomal fragmentation (Scoltock et al., 2000). Although the picture is far from complete, CLU protein has a signal sequence targeting nucleus (nuclear localization site, NLS), nuclear export sequence (NES), and also dinucleotide binding site (DBS). Accordingly, pnCLU is movable to nucleus although nCLU depends on deficiency of intracellular ionized calcium (Caccamo *et al.*, 2005; Pajak and Orzechowski, 2007a) and perhaps additional modifications (Shannan *et al.*, 2006b). The alternative scenario for CLU nuclear translocation was reported by O'Sullivan *et al.* (2003) who observed, that uncleaved, nonglycosylated, disulfide-linked CLU isoform targets nucleus in MCF-7 cells dying from apoptosis. Evidently, proapoptotic factors led to disruption of Golgi apparatus and incomplete posttranslational modifications of CLU. Surprisingly, mutation in NLS did not affect the nuclear accumulation of this form of CLU. The authors conclude that CLU is neither pro- nor antiapoptotic; nonetheless it interacts with Ku70, the protein partner of DNA-PK responsible for activating double-stranded DNA repair process (O'Sullivan *et al.*, 2003). The more nCLU, the less Ku70 for activation of DNA repair and apoptosis may proceed even in the absence of p53 (Chen *et al.*, 2004; Yang *et al.*, 2000). Conversely, secretion of sCLU shelters cells from apoptosis by scavenging the debris and phosphatidylserine while additionally stimulates *CLU* gene expression (Bach *et al.*, 2001). The role of Ca^{2+} in CLU activity seems to be cell-type specific, where deficiency of intracellular Ca^{2+} is required to halt CLU, shift it to nucleus, and to trigger apoptosis at least in prostate or colorectal cancer cells (Araki *et al.*, 2005; Caccamo *et al.*, 2005, 2006; Pajak and Orzechowski, 2007a). Taken together, diverse functions of CLU isoforms, or how they manage DNA integrity could be explained by intracellular setting where the same protein if secreted is antiapoptotic while nuclear isoform promotes cell death (Pajak and Orzechowski, 2006).

XI. CONCLUSIONS

Although CLU protein was discovered in testis more than quarter of century ago (Fritz *et al.*, 1983) its functions remain ambiguous, not clearly understood, or even contradictory. We here report a putative dynamic system based on Ca^{2+} oscillations that determine CLU anti- or proapoptotic characteristic. The presented observations suggest that Ca^{2+} fluxes act to either relay CLU secretion or grasp CLU for nuclear translocation. Consistent with a central role of Ca^{2+} in the final CLU fate, the increase in Ca^{2+} force out sCLU whereas drop of ionized Ca^{2+} facilitates tumor cells suicide by nCLU. Molecular switches regulated by Ca^{2+} signal (calnexin, calreticulin, calcineurin) and respective reciprocal hierarchy are also attractive pieces to elucidate any existing links between CLU and Ca^{2+}-dependent regulations.

Significant role in these events is played by intracellular and/or PMCCs including TRPM family of proteins. Although causality was not definitely established by use of calcium channel blockers and pharmacological probes, several lines of evidence support the notion that CLU extra- and intracellular translocations are modulated by Ca^{2+} signal. CLU is the unique chaperone that also plays a part in response to DNA damage. Although our current knowledge of the role of Ca^{2+} in CLU-dependent input to DNA repair, apoptosis, and cell cycle control highlights the relevance of these mechanisms both for carcinogenesis and progression of cancer, a lot more has to be done. Conflicting data relating Ca^{2+}/CLU contribution to the risk of cancer point to future studies to investigate whether calcium homeostasis and CLU are suitable targets for cancer prevention or therapy.

ACKNOWLEDGMENTS

This work was supported by grant No. N312 012 32/0761 from the Ministry of Higher Education in Poland. We apologize to our colleagues whose work could not be cited owing to space limitations. We gratefully acknowledge Prof. Saverio Bettuzzi, University of Parma, Italy, for his critical comments to the manuscript.

REFERENCES

Abed, E., and Moreau, R. (2009). Importance of melastatin-like transient receptor potential 7 and magnesium in the stimulation of osteoblast proliferation and migration by PDGF. *Am. J. Physiol. Cell Physiol.* **297**, C360–C368.

Ammar, H., and Closset, J. L. (2008). Clusterin activates survival through the phosphatidylinositol 3-kinase/Akt pathway. *J. Biol. Chem.* **283**, 12851–12861.

Araki, S., Israel, S., Leskov, K. S., Criswell, T. L., Beman, M., Klokov, D. Y., Sampalth, L., Reinicke, K. E., Cataldo, E., Mayo, L. D., and Boothman, D. A. (2005). Clusterin protein: Stress-inducible polypeptides with proposed functions in multiple organ dysfunction. *BJR Suppl.* **27**, 106–113.

Bach, U. C., Baiersdorfer, M., Klock, G., Cattaruzza, M., Post, A., and Koch-Brandt, C. (2001). Apoptotic cell debris and phosphatidylserine-containing lipid vesicles induce apolipoprotein J (clusterin) gene expression in vital fibroblasts. *Exp. Cell Res.* **265**, 11–20.

Balantinou, E., Trougakos, I. P., Chondrogianni, N., Margaritis, L. H., and Gonos, E. S. (2009). Transcriptional and posttranslational regulation of clusterin by the two main cellular proteolytic pathways. *Free Radic. Biol. Med.* **46**, 1267–1274.

Barbiero, G., Munaron, L., Antoniotti, S., Baccino, F. M., Bonelli, G., and Lovisolo, D. (1995). Role of mitogen-induced calcium influx in the control of the cell cycle in Balb-c 3T3 fibroblasts. *Cell Calcium* **18**, 542–556.

Barone, F., Aguanno, S., D-Alessio, A., and D'Agostino, A. (2003). Sertoli cell modulates MAA-induced apoptosis of germ cells throughout voltage-operated calcium channels. *FASEB J.* **18**, 353–364.

Barone, F., Aguanno, S., and D'Agostino, A. (2005). Modulation of MAA-induced apoptosis in male germ cells: Role of Sertoli cell P/Q-type calcium channels. *BMC Reprod. Biol. Endocrinol.* **3**, 13–22.

Batsi, C., Kontargiris, E., Markopoulou, S., Trougakos, I., Gonos, E. S., and Kolettas, E. (2008). Vanadium-induced apoptosis of HaCaT cells is mediated by *c-FOS* and involves up-regulation of nuclear clusterin/apolipoprotein J. In: *5th Clusterin/Apolipoprotein J (CLU) Workshop Spetses Island, Greece*, 2–5 June 2008 Book of abstracts 16.

Bettuzzi, S., Scorcioni, F., Astancolle, S., Davalli, P., Scaltriti, M., and Corti, A. (2002). Clusterin (SGP-2) transient overexpression decreases proliferation rate of SV40-immortalizaed human prostate epithelial cells by slowing down cell cycle progression. *Oncogene* **21**, 4328–4344.

Boynton, A. L., Whitefield, J. F., Isaacs, R. J., and Morton, H. J. (1974). Control of 3T3 cell proliferation by calcium. *In Vitro* **10**, 12.

Caccamo, A. E., Scaltriti, M., Caporali, A., D'Arca, D., Scorcioni, F., Candiano, G., Mangiola, M., and Bettuzzi, S. (2003). Nuclear translocation of a clusterin isoform is associated with induction of anoikis in SV-40-immortalized human prostate epithelial cells. *Ann. N. Y. Acad. Sci.* **1010**, 514–519.

Caccamo, A. E., Scaltriti, M., Caporali, A., D'Arca, D., Scorcioni, F., Astancolle, S., Mangiola, M., and Bettuzzi, S. (2004). Cell detachment and apoptosis induction of immortalized human prostate epithelial cells are associated with early accumulation of a 45 kDa nuclear isoform of clusterin. *Biochem. J.* **382**, 157–168.

Caccamo, A. E., Scaltriti, M., Caporali, A., D'Arca, D., Corti, A., Corvetta, D., Sala, A., and Bettuzzi, S. (2005). Ca^{2+} depletion induces nuclear clusterin, a novel effector of apoptosis in immortalized human prostate cells. *Cell Death Differ.* **12**, 101–104.

Caccamo, A. E., Desenzani, S., Belloni, L., Borghetti, A. F., and Bettuzzi, S. (2006). Nuclear clusterin accumulation during heat shock response: Implications for cell survival and thermotolerance induction in immortalized and prostate cancer cells. *J. Cell. Physiol.* **207**, 208–219.

Catterall, W. A. (2000). Structure and regulation of voltage-gated Ca^{2+} channels. *Annu. Rev. Dev. Biol.* **16**, 521–555.

Chayka, O., Corvetta, D., Dews, M., Caccamo, A. E., Piotrowska, I., Santilli, G., Gibson, S., Sebire, N. J., Himoudi, N., Hogarty, M. D., Anderson, J., Bettuzzi, S., et al. (2009). Clusterin a haploinsufficient tumor suppressor gene in neuroblastomas. *J. Natl. Cancer Inst.* **101**, 663–677.

Chen, T., Turner, J., McCarthy, S., Scaltriti, M., Bettuzzi, S., and Yeatman, T. (2004). Clusterin-mediated apoptosis is regulated by adenomatous polyposis coli and is p21 dependent but p53 independent. *Cancer Res.* **64**, 7412–7419.

Chiesa, R., Angeretti, N., Del Bo, R., Lucca, E., Munna, E., and Forioni, G. (1998). Extracellular calcium deprivation in astrocytes: regulation of mRNA expression and apoptosis. *J. Neurochem.* **70**, 1474–1483.

Chin, E. R., Olson, E. N., Richardson, J. A., Yang, Q., Humphries, C., Shelton, J. M., Wu, H., Zhu, W., Bassel-Duby, R., and Williams, R. S. (1998). A calcium-dependent transcriptional pathway controls muscle fiber type. *Genes Dev.* **12**, 2499–2509.

Criswell, T., Klokov, K., and Boothman, D. (2003). Transcriptional repression of clusterin by the p53 tumor suppressor protein. *Cancer Biol. Ther.* **2**, 25–31.

Denzel, A., Molinari, M., Trigueros, C., Martin, J. E., Velmurgan, S., Brown, S., Stamp, G., and Owen, M. J. (2002). Early postnatal death and motor disorders in mice congenitally deficient in calnexin expression. *Mol. Cell. Biol.* **22**, 7398–7404.

El Hiani, Y., Ahidouch, A., Lehen'kyi, V., Hague, F., Gouilleux, F., Mentaverri, R., Kamel, S., Lassoned, K., Brule, G., and Ouadid-Ahidouch, H. (2009). Extracellular signal-regulated kinases 1 and 2 and TRPC1 channels are required for calcium sensing receptor-stimulated MCF-7 breast cancer cell proliferation. *Cell Physiol. Biochem.* **23**, 335–346.

Estacion, M., and Mordan, L. J. (1993). Competence induction by PDGF requires sustained calcium influx by mechanism distinct from storage-dependent calcium influx. *Cell Calcium* **14**, 439–454.

Fink, T. M., Zimmer, M., Tschopp, J., Etienne, J., Jenne, D. E., and Lichter, P. (1993). Human clusterin (CLI) maps to 8p21 in proximity to the lipoprotein lipase (LPL) gene. *Genomics* **16**, 526–528.

Fragale, A., Aguanno, S., Kemp, M., Reeves, M., Price, K., Beattie, R., Craig, P., Volsen, P., Sher, E., and D'Agostino, A. (2000). Identification and cellular localization of voltage-operated calcium channels in mature rat testis. *Mol. Cell. Endocrinol.* **162**, 25–33.

Fritz, I. B., Burdzy, K., Setchell, B., and Blaschuk, O. (1983). Ram rete testis fluid contains a protein (clusterin) which influences cell–cell interactions *in vitro. Biol. Reprod.* **28**, 1173–1188.

Guo, L., Nakamura, K., Lynch, J., Opas, M., Olson, E. N., Agellon, L. B., and Michalak, M. (2002). Cardiac-specific expression of calcineurin reverses embryonic lethality in calreticulin-deficient mouse. *J. Biol. Chem.* **277**, 50776–50779.

Han, B. H., DeMattos, R. B., Dugan, L. L., Kim-Han, J. S., Brendza, R. P., Fryer, J. D., Kierson, M., Cirrito, J., Quick, K., Harmony, J. A. K., Aronow, B. J., and Holtzman, D. M. (2001). Clusterin contributes to caspase-3-independent brain injury following neonatal hypoxia-ischemia. *Nat. Med.* **7**, 338–343.

Illario, M., Cavallo, A. L., Bayer, U. K., Di Matola, T., Fenzi, G., Rossi, G., and Vitale, M. (2008). Calcium/calmodulin-dependent protein kinase II binds to Raf-1 and modulates integrin-stimulated ERK activation. *J. Biol. Chem.* **278**, 45101–45108.

July, L. L., Beraldi, E., So, A., Fazli, L., Evans, K., English, J. C., and Gleave, M. E. (2004). Nucleotide-based therapies targeting clusterin chemosensitize human lung adenocarcinoma cells both *in vitro* and *in vivo. Mol. Cancer Ther.* **3**, 223–232.

Kaisman-Elbaz, T., Sekler, I., Fishman, D., Karol, N., Forberg, M., Kahn, N., Hershfinkel, M., and Silverman, W. F. (2009). Cell death induced by zinc and cadmium is mediated by clusterin in cultured mouse seminiferous tubules. *J. Cell Physiol.* **220**, 222–229.

Kay, A. R. (2004). Detecting and minimizing zinc contamination in physiological solutions. *BMC Physiol* **4**, 4.

Khundmiri, S. J., Metzler, M. A., Ameen, M., Amin, V., Rane, M. J., and Delamere, A. (2006). Ouabain induces cell proliferation through calcium-dependent phosphorylation of Akt (protein kinase B) in opossum kidney proximal tubule cells. *Am. J. Physiol. Cell Physiol.* **291**, C1247–C1257.

Kojima, I., Matsunaga, H., Kurokawa, K., Ogata, E., and Nishimoto, I. (1988). Calcium influx: an intracellular message of the mitogenic action of insulin-like growth factor-I. *J. Biol. Chem.* **263**, 16561–16567.

Lallet-Daher, H., Roudbaraki, M., Bavencoffee, A., Mariot, P., Gackiere, F., Bideaux, G., Urbain, R., Gosset, P., Delcourt, P., Fleurisse, L., Slomianny, C., Dewailly, E., *et al.* (2009). Intermediate-conductance Ca^{2+}-activated K^+ channels (IKCa1) regulate human prostate cancer cell proliferation through a close control of calcium entry. *Oncogene* **28**, 1792–1806.

Leskov, K. S., Klokov, D. Y., Li, J., Kinsella, T. J., and Boothman, D. A. (2003). Synthesis and functional analyses of nuclear clusterin, a cell death protein. *J. Biol. Chem.* **278**, 11590–11600.

Liao, J., Schneider, A., Datta, N. S., and McKauley, L. K. (2006). Extracellular calcium as a candidate mediator of prostate cancer skeletal metastasis. *Cancer Res.* **77**, 9065–9073.

Lin, K. I., Chattopadhyay, N., Bai, M., Alvarez, R., Dang, C. V., Baraban, J. M., Brown, E. M., and Ratan, R. R. (1998). Elevated extracellular calcium can prevent apoptosis via the calcium-sensing receptor. *Biochem. Biophys. Res. Commun.* **249**, 325–331.

Liu, T., Liu, P. Y., Tee, A. E. L., Haber, M., Norris, M. D., Gleave, M. E., and Marshall, G. M. (2009). Over-expression of clusterin is a resistance factor to the anti-cancer effect of histone deacetylase inhibitors. *Eur. J. Cancer* **45**, 1846–1854.

Losh, A., and Koch-Brandt, C. (1995). Dithiothreitol treatment of Madin-Darby canine kidney cells reversibly blocks export from the endoplasmic reticulum but does not affect vectorial targeting of secretory proteins. *J. Biol. Chem.* **270**, 11543–11548.

Lu, J., McKinsey, T. A., Nicol, R. L., and Olson, E. N. (2000). Signal-dependent activation of the MEF2 transcription factor by dissociation from histone deacetylases. *Proc. Natl. Acad. Sci. USA* **97**, 4070–4075.

Lynch, J., and Michalak, M. (2003). Calreticulin is an upstream regulator of calcineurin. *Bichem. Biophys. Res. Commun.* **311**, 1173–1179.

Ma, H., Matsunaga, H., Li, B., Schieffer, B., Marrero, M. B., and Ling, B. N. (1996). Ca^{2+} channel activation by PDGF-induced tyrosine phosphorylation and Ras guanine triphosphate binding proteins in rat glomerular mesangial cells. *J. Clin. Invest.* **97**, 2332–2341.

Mailland, M., Waechli, R., Ruat, M., Boddeke, H. G. W., and Seuwen, K. (1997). Stimulation of cell proliferation by calcium and a calcimimetic compound. *Endocrinology* **138**, 3601–3605.

Mallory, J. C., Crudden, G., Oliva, A., Saunders, C., Stromberg, A., and Craven, R. J. (2005). A novel group of genes regulates susceptibility to antineoplastic drugs in highly tumorigenic breast cancer cells. *Mol. Pharmacol.* **68**, 1747–1756.

Mao, Z., Bonni, A., Xia, F., Nadal-Vicens, M., and Greenberg, M. E. (1999). Neuronal activity-dependent cell survival mediated by transcription factor MEF2. *Science* **285**, 785–790.

Marin, A., Lopez de Cerain, A., Hamilton, E., Lewis, A. D., Martinez-Penuela, J. M., Idoate, M. A., and Bello, J. (1997). DT-diaphorase and cytochrome B5 reductase in human lung and breast tumours. *Br J. Cancer* **76**, 923–929.

Markopoulou, S., Kontargiris, E., Batsi, C., Tzavaras, T., Trougakos, I., Boothman, D. A., Gonos, E. S., and Kolettas, E. (2009). Vanadium-induced apoptosis of HaCaT cells is mediated by c-Fos and involves nuclear accumulation of clusterin. *Febs J.* **276**, 3784–3799.

McKinsey, T. A., Zhang, C. L., and Olson, E. N. (2002). MEF2: A calcium-dependent regulator of cell division, differentiation and death. *Trends Biochem. Sci.* **59**, 271.

Mesaeli, N., Nakamura, K., Zvaritch, E., Dickie, P., Dziak, E., Krause, K. H., Opas, M., MacLennan, D. H., and Michalak, M. (1999). Calreticulin is essential for cardiac development. *J. Cell Biol.* **144**, 857–868.

Michalak, M., Groenendyk, J., Szabo, E., Gold, L. I., and Opas, M. (2009). Calreticulin, a multi-process calcium-buffering chaperone of the endoplasmic reticulum. *Biochem. J.* **417**, 651–666.

Miyake, H., Hara, I., Kamidono, S., Gleave, M. E., and Eto, H. (2003). Resistance to cytotoxic chemotherapy-induced apoptosis in human prostate cancer cells is associated with intracellular clusterin expression. *Oncol. Rep.* **10**, 469–473.

Molinari, M., Eriksson, K. K., Calanca, V., Galli, C., Creswell, P., Michalak, M., and Helenius, A. (2004). Contrasting functions of calreticulin and calnexin in glycoprotein folding and ER quality control. *Mol. Cell* **13**, 125–135.

Monaco, S., Illario, M., Rusciano, M. R., Gragnaniello, G., Di Spigna, G., Leggiero, E., Pastore, L., Fenzi, G., Rossi, E., and Vitale, M. (2009). Insulin stimulates fibroblast proliferation through calcium-calmodulin-dependent kinase II. *Cell Cycle* (epub ahead of print).

Munaron, L., Distasi, C., Carabelli, V., Baccino, F. M., Bonelli, G., and Lovisolo, D. (1995). Sustained calcium influx activated by basic fibroblast growth factor in Balb-c 3T3 fibroblasts. *J. Physiol.* **484**, 557–566.

Nagamine, K., Kudoh, J., Minoshima, S., Kawasaki, K., Asakawa, S., Ito, F., and Shimizu, N. (1998). Molecular cloning of a novel pytative Ca^{2+} channel protein (TRPC7) highly expressed in brain. *Genomics* **54**, 124–131.

Nizard, P., Tetley, S., Le Drean, Y., Watrin, T., Le Goff, P., Wilson, M. R., and Michel, D. (2007). Stress-induced retrotranslocation of clusterin/ApoJ into the cytosol. *Traffic* **8**, 554–565.

Ori, Y., Herman, M., Chagnac, A., Malachi, T., Gafter, U., and Korzets, A. (2005). Spontaneous DNA repair in human mononuclear cells is calcium-dependent. *Biochem. Biophys. Res. Commun.* **336**, 842–846.

Orzechowski, A., Jank, M., Gajkowska, B., Sadkowski, T., Godlewski, M.M, and Ostaszewski, P. (2003). Delineation of signalling pathway leading to antioxidant-dependent inhibition of dexamethasone-mediated muscle cell death. *J. Muscle Res. Cell Motil.* **24**, 33–53.

Orzechowski, A., Jank, M., Gajkowska, B., Sadkowski, T., and Godlewski, M. M. (2004). A novel antioxidant-inhibited dexamethasone-mediated and caspase-3 independent muscle cell death. *Ann. N. Y. Acad. Sci.* **1010**, S: 205–208.

O'Sullivan, J., Whyte, L., Drake, J., and Tenniswood, M. (2003). Alterations in the post-translational modifications and intracellular trafficking of clusterin in MCF-7 cells during apoptosis. *Cell Death Differ.* **10**, 914–927.

Pajak, B., and Orzechowski, A. (2006). Clusterin: The missing link in the calcium-dependent resistance of cancer cells to apoptogenic stimuli. *Postepy Hig. Med. Dosw.* **60**, 45–51.

Pajak, B., and Orzechowski, A. (2007a). Ethylenediaminetetraacetic acid affects subcellular expression of clusterin protein in human colon adenocarcinoma COLO 205 cell line. *Anti-Cancer Drugs* **18**, 55–63.

Pajak, B., and Orzechowski, A. (2007b). Sodium butyrate-dependent sensitization of human colon adenocarcinoma COLO 205 cells to TNF-α-induced apoptosis. *J. Physiol. Pharmacol.* **58**(Suppl. 3), 163–176.

Pajak, B., Gajkowska, B., and Orzechowski, A. (2009). Sodium butyrate sensitizes human colon adenocarcinoma COLO 205 cells to both intrinsic and TNF-α-dependent extrinsic apoptosis. *Apoptosis* **14**, 203–217.

Peppelenbosch, M. P., Tertoolen, L. G. J., den Hertog, J., and de Laat, S. W. (1992). Epidermal growth factor activates calcium channels by phospholipase $A_2/5$-lipooxygenase mediated leukotriene C_4 production. *Cell* **69**, 295–303.

Perraud, A. L., Schmitz, C., and Scherenberg, A. M. (2003). TRPM2 Ca^{2+} permeable cation channels: from gene to biological function. *Cell Calcium* **33**, 519–531.

Pinton, P., Giorgi, C., Siviero, R., Zecchini, E., and Rizutto, R. (2008). Calcium and apoptosis: ER-mitochondria Ca2+ transfer in the control of apoptosis. *Oncogene* **27**, 6407–6418.

Pu, Y., and Chang, D. C. (2001). Cytosolic Ca(2+) signal is involved in regulating UV-induced apoptosis in HeLa cells. *Biochem. Biophys. Res. Commun.* **23**, 84–89.

Pucci, S., Paola, M., Fabiola, S., David, B. A., and Luigi, S. G. (2009). Interleukin-6 affects cell death escaping mechanisms acting on Bax-Ku70-Clusterin interactions in human colon cancer progression. *Cell Cycle* **8**, 473–481.

Purrello, M., Bettuzzi, S., Di Pietro, C., Mirabile, E., Di Blasi, M., Rimini, R., Grzeschik, K. H., Ingletti, C., Corti, A., and Sichel, G. (1991). The gene for SP-40,40, human homolog of rat sulfated glycoprotein 2, rat clusterin, and rat testosterone-repressed prostate message 2, maps to chromosome 8. *Genomics* **10**, 151–156.

Ranney, M. K., Ahmed, I. S. A., Potts, K. R., and Craven, R. J. (2007). Multiple pathways regulating the anti-apoptotic protein clusterin in breast cancer. *Biochim. Biophys. Acta* **1772**, 1103–1111.

Rauhala, H. E., Porkka, K. P., Saramaki, O. R., Tammela, T. L., and Visakorpi, T. (2008). Clusterin is epigenetically regulated in prostate cancer. *Int. J. Cancer* **123**, 1601–1609.

Redondo, M., Tellez, T., Roldan, M. J., Serrano, A., Garcia-Aranda, M., Gleave, M. E., Hortas, M. L., and Morell, M. (2007). Anticlusterin treatment of breast cancer cells increases the sensitivities of chemotherapy and tamoxifen and counteracts the inhibitory action of dexamethasone on chemotherapy-induced cytotoxicity. *BMC Breast Cancer Res.* **9**, R86–R93.

Rizzi, F., Caccamo, A. E., Belloni, L., and Bettuzzi, S. (2009). Clusterin is a short half-life, poly-ubiquitinated protein, which controls the fate of prostate cancer cells. *J. Cell Physiol.* 219, 314–323.

Rose, J. K., and Doms, R. W. (1988). Regulation of protein export from the endoplasmic reticulum. *Ann. Rev. Cell Biol.* 4, 257–288.

Rowling, P. J. E., and Freedman, R. E. (1993). Folding, assembly, and posttranslational modification of proteins within the lumen of the endoplasmic reticulum. *Subcell. Biochem.* 21, 41–80.

Santilli, G., Aronow, B. J., and Sala, A. (2003). Essential requirement of apolipoprotein J (clusterin) signaling for IκB expression and regulation of NF-κB activity. *J. Biol. Chem.* 278, 38214–38219.

Scaltriti, M., Santamaria, A., Paciucci, R., and Bettuzzi, S. (2004). Intracellular clusterin induces G$_2$-M phase arrest and cell death in PC-3 prostate cancer cells. *Cancer Res.* 64, 6174–6182.

Scoltock, A. B., Bortner, C. D., Bird, G. St. J., Putney, J. W., and Cidlowski, J. A. (2000). A selective requirement for elevated calcium in DNA degradation, but not early events in anti-Fas-induced apoptosis. *J. Biol. Chem.* 275, 30586–30596.

Shannan, R., Seifert, M., Boothman, D. A., Tilgen, W., and Reichrath, J. (2006a). Clusterin and DNA repair: a new function in cancer for a key player in apoptosis and cell cycle control. *J. Mol. Histol.* 37, 183–188.

Shannan, B., Seifert, M., Leskov, K., Willis, J., Boothman, D., Tilgen, W., and Reichrath, J. (2006b). Challenge and promise: role of clusterin in pathogenesis, progression and therapy of cancer. *Cell Death Differ.* 13, 12–19.

Skinner, H. G., and Schwartz, G. G. (2009). A prospective study of total and ionized serum calcium and fatal prostate cancer. *Cancer Epidemiol. Biomarkers Prev.* 18, 575–578.

Stokes, A. H., Freeman, W. M., Mitchell, S. G., Burnette, T. A., Hellmann, G. M., and Vrana, K. E. (2002). Induction of GADD45 and GADD153 in neuroblastoma cells by dopamine-induced toxicity. *Neurotoxicology* 23, 675–684.

Tagliarino, C., Pink, J. J., Dubyak, G. R., Nieminen, A- L., and Boothman, D. A. (2001). Calcium is a key signaling molecule in β-lapachone-mediated cell death. *J. Biol. Chem.* 276, 19150–19159.

Tagliarino, C., Pink, J. J., Reinicke, K. E., Simmers, S. M., Wuerzberger-Davis, S. M., and Boothman, D. A. (2003). -calpain activation in β-lapachone-mediated apoptosis. *Cancer Biol. Ther.* 2, 141–152.

Taranta, A., Morena, A. R., Barbacci, E., and D'Agostino, A. (1997). Conotoxin-sensitive Ca^{2+} voltage-gated channels modulate protein secretion in cultured rat Sertoli cells. *Mol. Cell. Endocrinol.* 126, 117–123.

Taranta, A., Teti, A., Stefanini, M., and D'Agostino, A. (2000). Immediate cell signal induced by laminin in rat Sertoli cells. *Matrix Biol.* 19, 11–18.

Trougakos, I. P., and Gonos, E. S. (2004). Functional analysis of clusterin/apolipoprotein J in cellular death induced by severe genotoxic stress. *Ann. N. Y. Acad. Sci.* 1019, 206–210.

Trougakos, I. P., Lourda, M., Antonelou, M. H., Kletsas, D., Gorgoulis, V. G., Papassideri, I. S., Zou, Y., Margaritis, L. H., Boothman, D. A., and Gonos, E. S. (2009). Intracellular clusterin inhibits mitochondrial apoptosis by suppressing p53-activating stress signals and stabilizing the cytosolic Ku70-Bax protein complex. *Clin. Cancer Res.* 15, 48–59.

Tschopp, J., Jenne, D. E., Hertig, S., Preissner, K. T., Morgenstern, H., Sapino, A- P., and French, L. (1993). Human megakaryocytes express clusterin and package it without apolipoprotein A-1 into α-granules. *Blood* 82, 118–125.

Tsubuki, S., Saito, Y., Tomioka, M., Ito, H., and Kawashima, S. (1996). Differential inhibition of calpain and proteasome activities by peptidyl aldehydes of di-leucine and tri-leucine. *J. Biochem.* 119, 572–576.

Voets, T., Droogmans, G., Wissenbach, U., Janssens, A., Flockerzi, V., and Nilius, B. (2004). The principle of temperature-dependent gating in cold- and heat-sensitive TRP channels. *Nature* **430**, 748–754.

Wang, H- P., Pu, X- Y., and Wang, X- H. (2007). Distribution profiles of transient receptor potential melastatin-related and vanilloid-related channels in prostate tissue in rat. *Asian J. Androl.* **9**, 634–640.

Williams, D. B. (2005). Beyond lectins: The calnexin/careticulin chaperone system of the endoplasmic reticulum. *J. Cell Sci.* **119**, 615–623.

Yang, C- R., Leskov, K., Hosley-Eberlein, K., Criswell, T., Pink, J. J., Kinsella, T. J., and Boothman, D. A. (2000). Nuclear clusterin/XIP8, an x-ray-induced Ku70-binding protein that signals cell death. *Proc. Natl. Acad. Sci. USA* **97**, 5907–5912.

Youn, H- D., Sun, L., Prywes, R., and Liu, J. O. (1999). Apoptosis of T cells mediated by Ca^{2+}-induced release of the transcription factor MEF2. *Science* **286**, 790–793.

Zellweger, T., Chi, K., Miyake, H., Adomat, H., Kiyama, S., Skov, K., and Gleave, M. E. (2002). Enhanced radiation sensitivity in prostate cancer by inhibition of the cell survival protein clusterin. *Clin. Cancer Res.* **8**, 3276–3284.

Zhang, H., Kim, J. K., Edwards, C. A., Xu, Z., Taichman, R., and Wang, C- Y. (2005). Clusterin inhibits apoptosis by interacting with activated Bax. *Nat. Cell Biol.* **7**, 909–915.

Nuclear CLU (nCLU) and the Fate of the Cell

Saverio Bettuzzi and Federica Rizzi

*Dipartimento di Medicina Sperimentale, Sezione di
Biochimica, Biochimica Clinica e Biochimica dell'Esercizio
Fisico, Via Volturno 39-43100 Parma; and
Istituto Nazionale Biostrutture e Biosistemi
(I.N.B.B.), Rome, Italy*

The possible biological role played by Clusterin (CLU) has been puzzling researchers for a long time since its first discovery and characterization. CLU has been often
described as an "enigmatic" gene, a clear indication that too many aspects of this issue
have been obscure or difficult to interpret for long. The good news is that this is
certainly no longer true. Since the beginning, CLU was believed to play important
roles in nearly all most important biological phenomena. The diversity, sometime the

0065-230X/09 $35.00
DOI: 10.1016/S0065-230X(09)04005-6

contradictions, of its biological action is now likely explained by the existence of different protein products all generated by the same single copy *CLU* gene. The relatively recent discovery that CLU can be retained inside the cell and targeted to many intracellular sites and organelles, including the nucleus, provided us a very different view from that solely deriving from its possible role in the outer cellular environment. In particular, nuclear localization of CLU (nCLU) was found to trigger cell death in many systems.

In this chapter, a critical review of previous work will enable us to reinterpret old data and observations in the attempt to progressively unravelling the CLU "enigma" by considering its localization inside and outside the cell. The final picture would supposedly reconcile different or alternative hypothesis. Starting with an "historical" approach demonstrating that nCLU was right under our eyes since the beginning, up to the more recent contributions we will describe which stimuli would inhibit secretion and maturation of CLU leading at least one protein product to target the nucleus and kill the cell. A better understanding of this complex issue is not an easy work, considering the thoughtfulness in reviewing the existing literature and the known controversial reports. We hope that the information contained in this article will be useful for the reader to enlighten this field. © 2009 Elsevier Inc.

I. INTRODUCTION: WHERE IS CLU?

The discovery that Clusterin (CLU) can be retained inside the cell and targeted to many intracellular sites and organelles, including the nucleus, is rather recent. Before all this, CLU was known as an extracellular protein, heavily glycosylated, and expelled from the cell as secreted product into the outer cellular environment. Secreted CLU (sCLU) can be found in all the body fluids studied, in different amounts. The rather well-known description of the maturation process of CLU, as it can be found in many reviews in the literature, usually refers to the secreted protein product. At difference, we know very little about the structure of the intracellular products of CLU or the molecular mechanisms driving this process. But we know a bit more about which stimuli would inhibit secretion and maturation, causing CLU to target many intracellular sites and the nucleus. Very importantly, we also know about the biological effects caused by nuclear targeting of CLU, which will be our main topic in this chapter.

Although we know that important changes in the molecular weight, as revealed by Western blot analysis, are associated to intracellular processing and maturation of CLU, and that a 45–55 kDa protein band (the range is due to similar but not identical observations in different research reports) is usually associated to the detection of CLU inside the nucleus, we do not have definitive information about the structural differences between these protein products (please see also chapters "Introduction" and "Clusterin

(CLU): From one gene and two transcripts to many proteins"). Most importantly, we lack definitive structural information of CLU in relation to its intracellular localization. sCLU is known to be a very sticky protein outside the cell, where it can act as an extracellular chaperone (see chapters "The chaperone action of CLU and its putative role in quality control of extracellular protein folding" and "Cell protective functions of secretory CLU (sCLU)"). Probably for the same reason, when CLU is retained within the cell, it can be found nearly everywhere bound to organelles (mitochondria), membranes, cytoskeleton, nucleus, and many other cell structures. In this chapter, we will use the term sCLU when the most of CLU is fully maturated in the Golgi apparatus and exported from the cell as glycosylated protein. Under these conditions, usually the cytoplasm of the healthy cell is almost depleted of CLU, which can only be found in small amounts in vesicles and Golgi. At difference, the secretion of CLU can be completely abolished. Under these conditions, CLU is not detectable any longer in the cell-culture medium and the cells will get fully loaded with CLU. Now CLU can be mostly found in the cytoplasm of the cell, but also in the nucleus. The condition in which CLU is only present in the cytoplasm as well as that in which CLU is only present in the nucleus is very rare. The most common situation is the detection of an "intracellular" staining showing CLU both in the cytoplasm and in the nucleus. The detection of even a small amount of CLU in the nucleus of intact cells by appropriate methods is here defined as nuclear CLU (nCLU). Therefore, cytoplasm and nuclear localization may coexist. Importantly, we will see that when nCLU prevails, the cell is doomed to die very rapidly and disappear from the culture or the tissue. Detection of prevailing nCLU in living cell is not easy, and can only be done by taking advantage of a window of opportunity that can last very shortly.

Here, we will describe how and why the presence of even small amounts of nCLU will drive the fate of the cell. Detection of nCLU is here intended more as a functional condition rather than a structurally defined molecular form. This is why we will not use the term "isoform." At the moment we are not aware of definitive experimental data showing which structural domains are characteristics of each different protein product, as well as we do not exactly know which domains are necessary for nuclear targeting, since at least three nuclear localization signal (NLS), present in the human CLU amino acid sequence, have been identified but found to be unnecessary for nuclear targeting. Thus, while we have a sketchy idea about the structure of sCLU, we still do not know how nCLU looks like. Waiting for more data about that, we will focus on the biological effects of nCLU and their potential implication for the fate of the cell.

II. PRELIMINARY REMARKS BEFORE DEALING WITH nCLU: THE POSSIBLE ACTION OF sCLU AS DIFFERENTIATING AGENT

Although this topic will be specifically discussed in chapters "The chaperone action of CLU and its putative role in quality control of extracellular protein folding" and "Cell protective functions of secretory CLU (sCLU)," we would like to provide some preliminary remarks on the action of soluble sCLU because we consider these informations are instrumental to a better understanding of the topic of this chapter, that is, nCLU. Extracellular sCLU is often considered to be a survival factor. Programmed cell death is often associated to active cell cycling. Quiescence and differentiation are believed to be conditions rather refractory to apoptosis. Therefore, a prodifferentiating activity can be easily seen as prosurvival or antiapoptotic as well. This issue is relevant, because when CLU is actively exported from the cell as sCLU, nCLU is not detectable any longer inside the cell. And vice versa, in a Yin-Yang fashion. To our opinion, important experimental data were obtained in a pioneer work on differentiating smooth muscle cells (SMC). SMC grown to a high-density monolayer culture undergo a morphological transition and form multicellular nodules, resembling SMC present in the aortic media and in some atherosclerotic plaques. In an early observation, the process of nodule formation was found associated with the enhanced production of a secreted 38-kDa glycoprotein. After cloning and sequencing, the authors (Diemer *et al.*, 1992) demonstrated that the 1646-base pair cDNA, containing a single open reading frame encoding 446 amino acids, was actually homolog to the human complement cytolysis inhibitor (CLI), also called serum protein-40,40 (SP-40,40), or sulfated glycoprotein-2 (SGP-2); in other words, CLU. The level of expression of CLU mRNA increased as the cultures begin to form multilayered regions (Diemer *et al.*, 1992). Later on, these data were further confirmed when SMC cultures were studied after seeding on a preformed extracellular matrix composed of Matrigel. Under these conditions, SMC formed nodules within 24 h. Cultures seeded on a collagen gel formed nodules slightly later, at 48–72 h. Surprisingly, it was found that Matrigel contains sCLU as a major component. The authors proposed that endogenous CLU supports the rapid formation of nodules. In fact, nodule formation in SMC cultures growing on collagen gel is inhibited by the addition of anti-CLU antibody. The collagen gel does not contain CLU, but it was found to induce CLU expression in SMC. Also in this case, formation of nodules by SMC growing on collagen gels was inhibited by administration of anti-CLU antibody or by preincubation with sCLU purified from plasma. These results suggest that sCLU has a functional role in modulation of SMC proliferation and differentiation (Thomas-Salgar and Millis, 1994). The investigation continued later. To directly test the hypothesis that sCLU was directly responsible of SMC phenotypic modulation, cultured vascular SMC were stably transfected with an expression plasmid for manipulation of the level of expression of CLU. Twenty-four clones were selected and characterized for CLU expression and culture morphology. Clone SM-CLU18AS, expressing and secreting high levels of CLU, formed multicellular nodules, while clone SM-CLU13AS, expressing low level of CLU, was unable to form nodules even in the presence of a preformed collagen gel. Importantly, CLU-negative SM-CLU13AS were still capable to form nodules when exogenous CLU (i.e., sCLU) was added to the system. SM-CLU13AS formed nodules when cultured in Matrigel and in the presence of sCLU-containing conditioned media prepared from nodular SMC or SM-CLU18AS cultures. These results demonstrate that CLU is required for formation of SMC nodules, suggesting that CLU may play an important role in the morphological transition of SMC and their reorganization in the tissue (Moulson and Millis, 1999).

In other experiments, butyrate, a powerful antiproliferative agent, a strong promoter of cell differentiation, and an inducer of apoptosis, was used to inhibit vascular smooth muscle cells (VSMC) proliferation at physiological concentrations. Then expression of genes differentially related to growth and differentiation was studied. Results indicated that butyrate-inhibited VSMC proliferation was related to the downregulation of genes encoding positive regulators of cell growth. At the same time, upregulation of negative regulators of growth or differentiation inducers was detected. Among the downregulated genes, PCNA, retinoblastoma susceptibility-related protein p130 (pRb), cell division control protein 2 homolog (cdc2), cyclin B1, cell division control protein 20 homolog (p55cdc), high mobility group (HMG) 1 and 2, and several others were found. On the other hand, upregulated genes (i.e., negative regulators of growth or differentiation inducers) included cyclin D1, p21WAF1, p14INK4B/p15INK5B, inhibitor of DNA binding 1 (ID1), noticeably CLU, and others (Ranganna et al., 2003). Thus, CLU was characterized as a negative regulator of growth (or a differentiation inducer) in this system. In addition to this, sCLU was then found to be a potent inhibitor of VSMC migration, adhesion, and proliferation. In a recent work, CLU was identified as upregulated in distal anastomotic intimal hyperplasia after prosthetic arterial grafting. The effect of sCLU on VSMC migration was studied using a microchemotaxis chamber. In this experimental model, administration of sCLU caused inhibition of migration, adhesion, and incorporation of thymidine in VSMC. By means of microarray analysis, it was found that exposure of VSMC to sCLU caused upregulation of interleukin-8 and endothelin-1, thus suggesting that the possible genetic targets of sCLU are linked to cell senescence and differentiation (Sivamurthy *et al.*, 2001). A strong inhibitory effect of CLU on cell adhesion and migration and a clear link with anoikis-death was then confirmed also in prostate epithelial cells.

III. A BRIEF HISTORY OF nCLU

What may appear novel, actually it is not. In this chapter, we will show the reader that the existence of nCLU was hypothesized and somehow known long time ago. Clear evidences about nCLU date 1996. Since then, a critical review of the literature shows that, in the time-frame 1996–2002, about 25 papers referring to nCLU have been published. Later, from 2003 to June 2009, they were twice as much (about 53), demonstrating a growing interest concerning this issue. So, the new actor in the play, nCLU, is making the story more complex but also more interesting and challenging. It appears rationale to believe that the biological role of CLU outside and inside the cell is possibly very different. We have produced experimental data suggesting a very important role of CLU at deciding the final fate of the cell as a function of its localization inside or outside the cell. Understanding this scenario would enable us to build a paradigm, in which CLU is an example about how the merging of information from different fields, laboratories, and research teams may render an integrated view depicting a new scenario and suggesting new hypothesis capable to resolve apparent contradictions. One example for all: a typical contradiction found in the literature is the link between CLU and cell death. Is it CLU pro- or antiapoptotic? Does CLU plays any important role in cell death? Well, the answer "yes" to the second question may not be so straightforward. Considering that sCLU is present at high concentration in blood (100 μg/ml) and many fluids and secretions (1–15 μg/ml), we may speculate that, if the secreted form is proapoptotic, the most of the cells in the body would die. But even the reverse is hard to believe: if the secreted form is antiapoptotic, the most of the biological events requiring programmed cell death, which are absolutely necessary for an healthy body, would never occur. Thus, the story must be more complicated. For instance, many other players must be considered, because the possible biological effect observed in the end will be very likely the result from the action of many proteins, and not only CLU.

Before getting more deeply into these considerations, we need to know about how nCLU was found and its possible association with fundamental biological events such as apoptosis, proliferation, regeneration, and differentiation. The data here presented are derived from results obtained in different experimental models. In most of the cases reported, information was obtained in epithelial or neuroepithelial cells systems, sometimes in stromal cells.

IV. CLU EXPRESSION AND LOCALIZATION IN THE REGRESSING RAT VENTRAL PROSTATE

After the first isolation and characterization of extracellular CLU protein (sCLU) in ram rete testis fluid in 1983 (Blaschuk *et al.*, 1983; Fritz *et al.*, 1983), a truncated cDNA with no correspondent protein was cloned in the regressing rat ventral prostate following surgical castration and named testosterone repressed prostate message 2 (TRPM2) (Montpetit *et al.*, 1986). In 1989, we independently completed the cDNA cloning and sequencing of the most highly upregulated mRNA in the same experimental system (Bettuzzi *et al.*, 1989). After sequence comparison, we found a complete homology of our 1.7 kb cDNA to full-length SGP-2 cDNA. At the same time, we found that the truncated cDNA named TRPM2 was also identical to SGP-2. It was only 1 year before when it was found that SGP-2 was identical to CLU, a serum protein involved in aggregation of heterologous eritrocytes (Cheng *et al.*, 1988). Therefore, the circle was closed, and different research teams found that the same cDNA/protein was implicated in different phenomena at different body sites, and regulated in a very complex way not only by androgens but also by estrogens and other growth factors. In fact, it is important to know from the beginning that SGP-2/TRPM2 was also upregulated in the female rat uterus following ovariectomy (Bettuzzi *et al.*, 1989), thus CLU expression was not only regulated by androgens but also by estrogens at least. The common trait of these experimental models is that massive cell death is always induced following androgen or estrogens withdrawal in target organs such as prostate and uterus. The hypothesis that CLU expression may have much to do with cell death was a rather obvious consequence. When reliable antibodies against CLU were available, we studied the tissue distribution of CLU in the rat prostate during the time-course of androgen ablation by surgical removal of the testis. As shown in Fig. 1, we detected a wave of protein accumulation perfectly overlapping the time-frame induction of apoptotic death. Typically, the expression of *CLU* gene is induced right after castration reaching its highest level on the 4th day, then the protein products continue to accumulate for 3–4 more days. Later, the staining for CLU start to decrease levelling down to basal levels at 7–8 days

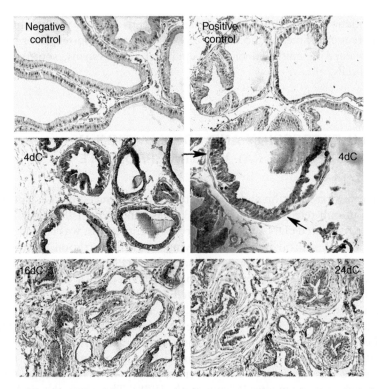

Fig. 1 Immunohistochemistry analysis of CLU expression showing the time-course induction of CLU overexpression and nCLU staining during the involution of the rat prostate gland following surgical removal of the testis. Positive control: prostate of an intact rat; 4dC, rat prostate 4 days after castration (left, 20×; right, 40×); 16dC, rat prostate 16 days after castration; 24dC, rat prostate 24 days after castration. Please note that nuclear staining of CLU (nCLU) is clearly evident in some basal cells at 4dC (arrows; 40×). The decrease in CLU nuclear staining occurring later after castration is probably due to the disappearance of cells in which nCLU has triggered cell death. All sections were counterstained with hematoxylin. (See Page 6 in Color Section at the back of the book.)

after castration. The time-course of CLU mRNA induction is identical by Northern blot analysis (Bettuzzi *et al.*, 1989).

Interestingly at 4 days after castration, when the peak of both CLU overexpression and cell death was reached (Bettuzzi *et al.*, 1989), a clear nuclear signal was easily detectable in basal cells of the glandular epithelia. The decrease in CLU nuclear staining occurring at later stages after castration is probably due to the disappearance of cells in which nCLU has triggered cell death. For what we know, nCLU induces anoikis-death very rapidly, so the target nCLU-positive cells are very likely difficult to detect because they rapidly disappear from the tissue *in vivo*. This experiment shows that, following the appropriate stimulus, CLU is targeted in the nucleus at appropriate time in the rat ventral prostate system. So, nCLU was already detectable long ago in one of the most typical experimental system where it was identified and described earlier.

V. CLU WAS FOUND POSITIVELY LINKED TO APOPTOSIS AND NEGATIVELY LINKED TO PROLIFERATION: THE EARLY YEARS (1993–1995)

The possible link between CLU expression and the fate of the cell was suggested long ago. A positive association with apoptotic death was found soon (Bettuzzi *et al.*, 1991). Many researchers have investigated on the same issue. In 1993–1995, nCLU was still not explicitly hypothesized, but several reports started to provide evidences about important biological effects of CLU. The recovery of the kidney after ischemic acute renal failure is an interesting model in which the role played by mitogenesis and dedifferentiation in the repair process can be investigated. In this model, various proteins implicated in mitogenesis, differentiation, and injury were identified. Proliferating cell nuclear antigen (PCNA), a specific marker for cell proliferation, was detected primarily in the S3 segment of the proximal tubule, with maximal expression at 2 days postischemia. Vimentin, normally present in mesenchymal cells but not in epithelial cells (it is considered a marker of dedifferentiation), was prominently expressed in the S3 segment at 2–5 days postischemia. Immunohistochemistry analysis showed that none of the cells positively stained with CLU antibodies were stained with PCNA or vimentin antibodies. Consistently, none of the PCNA or vimentin-positive cells were expressing CLU. Thus, at sites where significant ischemic injury occurs, surviving cells positively expressing a set of markers indicating that they were undergoing mitogenesis postischemically (PCNA and vimentin) did not express CLU (Witzgall *et al.*, 1994). In this experimental model, *CLU* gene expression was found inversely associated to proliferation and vimentin expression.

On the opposite hand, CLU expression was found positively associated with the quiescence state of the cell. Involution of the mouse mammary gland is characterized by loss of secretory epithelial cells. This phenomenon is massively caused by induction of programmed cell death. Programmed cell death in the mammary gland is associated with the expression of the growth arrest gene Gas-1, a marker of the quiescence state of cells. The irreversible phase of involution, characterized by internucleosomal DNA fragmentation, is preceded by nuclear activation of protein kinase A and transcription factor activator protein 1. Both events can be prevented by lactogenic hormone treatment in explant cultures derived from mammary tissue at lactation. In this system, the increase in activator protein 1 expression was found associated with the epithelial expression of CLU (SGP-2). CLU was suggested to be a potential target gene of activator protein 1 (Jaggi *et al.*, 1996). The strict association between inhibition of *CLU* gene expression and stimulation of cell proliferation was previously found in human lymphocytes synchronized in the quiescence state and then stimulated to proliferation by administration of phytohemagglutinins (PHA) (Grassilli *et al.*, 1991). Moreover, the close link between CLU and Gas-1 expression was later confirmed in another experimental system in which dermal stromal cells have been synchronized by serum starvation and then stimulated to growth. During the G_0 phase of the cell cycle, CLU expression reached the maximum level together with Gas-1, while stimulation of proliferation again inhibited *CLU* gene expression (Bettuzzi *et al.*, 2002) and Gas-1, as expected.

Tissue regeneration is another interesting model in which CLU expression can be studied and its possible function(s) challenged. Renal artery stenosis was used to induce unilateral noninfarctive renal atrophy (Gobe *et al.*, 1995). After several weeks, mitosis was at normal levels and atrophic kidney cells showed minimal apoptosis or inflammatory response. At this stage, regeneration of atrophic kidneys was stimulated by removal of the contra-lateral healthy kidneys. The regrowth response was very rapid, involving renal hyperplasia rather than hypertrophy. In the first 24 h of acute regenerative phase, CLU expression was markedly increased, decreasing later on to untraceable levels by 5 days of regeneration. Cell proliferation peaked at 3–5 days of regeneration and was localized in dedifferentiated tubules. CLU mRNA was actually found in dedifferentiated tubules, more precisely localized in dilated or collapsed atrophic tubules that had lost the identifying surface structures of normal tubular epithelium, and in the periphery of some blood vessel walls. Despite the regenerative stimulus, a transient but marked increase in apoptotic cell death in atrophic tubules in the first 24 h of regeneration was detected. Experimental data provided evidences of a temporal association between increased *CLU* gene expression and apoptosis. *In situ* localization showed CLU mRNA accumulated in apparently viable as well as apoptotic cells in the epithelium of tubules. CLU mRNA expression was rarely occurring in epithelial cells present in foci of nonatrophic (nondedifferentiated) nephrons that responded to the regenerative stimulus by cellular hypertrophy (Gobe *et al.*, 1995). In this model, *CLU* gene expression was clearly associated to the wave of apoptotic cell death rapidly occurring after the regenerative stimulus, and not to hyperproliferation.

VI. CLU AND FOLLICULAR ATRESIA, ANOTHER MODEL OF TISSUE INVOLUTION LINKED TO APOPTOSIS

Hypophysectomy is known to induce involution of follicles. Hypophysectomy-induced ovarian follicular atresia is an apoptotic process. Also in this system changes in the steady-state levels of ovarian CLU mRNA were found. Nuclear condensation, cytoplasmic shrinkage, and production of apoptotic bodies at all levels of the granulosa cell layer was demonstrated by electron microscopic analysis of the degenerating cells of atretic follicles. Upregulation of ovarian CLU gene expression was specifically seen in theca-interstitial cells, an effect that was prevented by the concurrent administration of FSH (an established antiatretic) or pregnant mare serum gonadotropin (PMSG) (Hurwitz et al., 1996).

VII. CLU EXPRESSION WAS FOUND ASSOCIATED TO NONAPOPTOTIC CELL DEATH

It is well known that the death of a cell is not only due to apoptosis. To make short a long story, beside by necrosis, a cell can die by other programmed cell death modalities already described, including autophagy. Cell death is more prominent in blastocysts from diabetic rats than control embryos. In this model, increased nuclear fragmentation occurs mainly in the inner cell mass. In addition, terminal transferase-mediated dUTP nick end labeling (TUNEL) demonstrated an increase in the incidence of non-fragmented DNA-damaged nuclei in these blastocysts. By quantitative RT-PCR, an increase in CLU mRNA levels was found in blastocysts from diabetic rats. In situ hybridization showed that only about half the cells expressing CLU mRNA had nuclear fragmentation. In vitro administration of high D-glucose increased nuclear fragmentation and TUNEL labeling, as well as CLU transcription. Tumor necrosis factor-alpha (TNF-α), a cytokine upregulated in the diabetic uterus, did not induce nuclear fragmentation nor CLU expression, but increased the incidence of TUNEL-positive nuclei, suggesting that CLU expression may be related to cell death not linked to DNA fragmentation (Pampfer et al., 1997). Peroxisome proliferator-activated receptor-gamma (PPAR-gamma) is a ligand-activated transcription factor that controls growth, differentiation, and inflammation in different tissues. AR42J cells were treated with troglitazone, a ligand of PPAR-gamma. Induction of apoptosis by troglitazone was evaluated by assessing cell viability, DNA fragmentation as well as by flow cytometry. Troglitazone induced the expression of pancreatitis-associated protein-1 and CLU mRNAs. Dose-dependent troglitazone-induced apoptosis was not blocked by inhibitors of caspases (Masamune *et al.*, 2002). Interestingly, CLU expression was again found associated to cell death even if not related to activation of caspases.

In another experiment (Zhang *et al.*, 1997), castrated male rats were treated with mimosine, a reversible S1-phase cell-cycle blocking drug. The effect of mimosine on prostate cell apoptosis caused by castration was studied. Following administration of a single dose of mimosine (25 mg/kg/day), internucleosomal DNA fragmentation associated with apoptosis was partially suppressed in the rat ventral prostate at all early time points analyzed postcastration (24, 48, and 72 h). Suppression was dose-dependent. Treatment with mimosine up to 150 mg/kg/day was sufficient to reduce the internucleosomal DNA fragmentation in the prostate by 90% at 72 h postcastration. Nevertheless, the treatment did not suppress the histological appearance of apoptotic bodies in the regressing ventral prostate. Thus, programmed cell death independent from DNA fragmentation was induced. Under these conditions, several mRNAs coding for apoptosis-associated genes such as bcl-2, p53, TGF-beta, and CLU were induced, to confirm that apoptosis was somehow involved. The fact that the apoptotic bodies found in mimosine-treated cells were lacking nuclear DNA fragmentation (Zhang *et al.*, 1997) suggested again that *CLU* gene overexpression may be associated to alternative cell death pathways.

VIII. CLU EXPRESSION WAS FOUND ASSOCIATED TO LESIONS IN CENTRAL AND PERIPHERAL NERVOUS SYSTEM

CLU gene expression was found associated to neurodegeneration in the brain (Laping *et al.*, 1994). Transforming growth factor-beta 1 (TGF-β1) was studied as a possible regulator of messenger RNAs synthesis in astrocytes and neurons. After hippocampal deafferentation by perforant path transaction, the mRNAs coding for tubulin alpha 1, glial fibrillary acidic protein, and CLU are increased. Because of the fact that TGF-β1 mRNA is increased following this lesion, a study aimed at identifying which mRNA could be induced by TGF-β1 alone was performed. To this aim, porcine TGF-β1 was infused into the lateral ventricle. The result was an increase in the mRNAs for tubulin alpha 1, glial fibrillary acidic protein, and CLU. This experimental model mimicked brain lesions, because 24 h after infusion in the ipsilateral hippocampus, TGF-β1 also increased several subsets of neuronal and astrocyte mRNAs coding for cytoskeletal proteins that are usually elevated in response to experimental lesions and Alzheimer's disease (Laping *et al.*, 1994).

CLU was found implicated in brain degeneration (Norman *et al.*, 1995). In the neurologically mutant mouse strain lurcher (Lc), heterozygous animals display autonomous degeneration of cerebellar Purkinje cells beginning in the second postnatal week. In this model, Lc Purkinje cells die by apoptosis, showing nuclear condensation, axon beading, and membrane blebbing. Experiments with *in situ* hybridization showed that CLU mRNA is not expressed at detectable levels in normal Purkinje cells, but only in Lc Purkinje cells just prior to their death. In the same model, expression of the Kv3.3b potassium channel, which marks the terminal phase of Purkinje cell differentiation, is also evident in Lc Purkinje cells prior to their death (Norman *et al.*, 1995). A similar association was found in peripheral nervous system degeneration. In fact, *CLU* gene expression was found altered (i.e., increased) in retinal degeneration (Jomary *et al.*, 1993). In the *rds* mutant mouse model, a model for the inherited human disease retinitis pigmentosa, the photoreceptors undergo a slow degeneration via apoptosis leading to eventual loss of the entire photoreceptor population. Scattered CLU-positive cells were found in the outer nuclear layer of dystrophic retinas. The increased level of CLU protein in the dystrophic retina was interpreted as associated to dying photoreceptor cells due to apoptosis. In this system, CLU protein and apoptotic nuclei were colocalized using brown and red coloured substrates, respectively. CLU was found colocalized with apoptotic nuclei both in the outer nuclear layer of *rds* mutant retinas as well as in the small intestine epithelial cells undergoing cell turnover and exfoliation (Agarwal *et al.*, 1996). These results are of high interest for at least two reasons: (i) overexpression of CLU was also observed in other neurodegenerative diseases such as Alzheimer's and Pick's disease; (ii) overexpression of CLU will be found later associated to APC expression in colon cells (Chen *et al.*, 2004), thus accounting for the finding that CLU was localized in small intestine epithelial cells undergoing turnover and exfoliation.

IX. INCREASED CLU EXPRESSION IS LINKED TO INHIBITION OF PROLIFERATION AND INDUCTION OF CELL DIFFERENTIATION

CLU expression was also studied during the regeneration of the pancreas in pancreatectomized rats, a known model of cell proliferation and differentiation (Min *et al.*, 2003). In this system, following the stimulus CLU was found transiently expressed in the differentiating acinar cells, but faded afterwards. As it was found before, CLU-positive cells were negative for PCNA and BrdU incorporation, whereas the most of epithelial cells in ductules of the regenerating pancreas showed increased proliferation activity. CLU was also found expressed in some endocrine cells of the regenerating tissue, in particular in the glucagon-secreting cells. Following a transient 2.5-fold increase in cell proliferation, transfection of CLU cDNA induced transformation of nondifferentiated duct cells into differentiated cells, displaying immunoreactivity for cytokeratin-20 (Min *et al.*, 2003). A negative association of CLU expression with cell proliferation and particularly to S-phase was previously observed about 1 year earlier (Bettuzzi *et al.*, 2002).

X. EARLY STUDIES ON TRANSLATIONAL UPREGULATION OF CLU DURING APOPTOSIS: THE ROLE OF NF1

Since the beginning of this story, CLU gene expression was found rapidly induced in early involution of the mouse mammary gland after weaning and the rat ventral prostate following surgical castration. A study was conducted in the search for DNA-binding proteins, regulatory elements on the promoter, and transcription factors involved in regulation of *CLU* gene expression. DNaseI footprints were used to this aim in the proximal mouse *CLU* gene promoter, leading to the identification and characterization of a twin nuclear factor 1 (NF1)-binding element at $-356/-309$. The study was conducted in both experimental models, prostate and mammary glands: nuclear extracts from 2-day regressing mouse mammary gland showed an enhanced footprint over the proximal NF1 element, while extracts from regressing prostate showed enhanced occupancy of both NF1-binding elements. Then, by EMSA and Western blot analysis, a 74-kDa NF1 protein whose expression is triggered in early involution in the mouse mammary gland was identified. This protein was found induced in the mammary system but not in the regressing rat ventral prostate. When cells were grown with laminin-rich extracellular matrix, apoptosis and the expression of the 74-kDa NF1 were both suppressed. A switch in the expression of different members of the NF1 transcription factors family was seen when the status of mammary epithelial cells changed from differentiated to involuted/apoptotic (Furlong *et al.*, 1996; Kane *et al.*, 2002). Unfortunately, this study did not provide hints about the regulation of CLU transcriptional activity in the regressing rat ventral prostate system.

XI. *CLU* GENE EXPRESSION AND CELL DEATH INDUCED BY VITAMIN D (CALCITRIOL) AND TAMOXIFEN (TAM): IMPLICATION IN CHEMIORESISTANCE

It is known that 1,25-dihydroxycholecalciferol D3 (1,25-(OH)2D3), the active metabolite of vitamin D, is a potent inhibitor of breast cancer cell growth both *in vivo* and *in vitro*. In a interesting early study (Welsh, 1994), it was shown that when MCF-7 cells are treated with 100 nM 1,25-(OH)2D3 apoptosis is induced as assessed by morphology (pyknotic nuclei, chromatin and cytoplasmic condensation, nuclear matrix protein reorganization). Cell death is triggered at 48 h after starting the treatment. The antiestrogen 4-hydroxytamoxifen (TAM) also induces apoptosis in MCF-7 cells. A combined treatment of both agent was attempted following the suggestion that TAM may act synergically and greatly enhance inhibition of cell growth caused by 1,25-(OH)2D3 alone. As a matter of fact the combined treatment with 1,25-(OH)2D3 and TAM enhanced the degree of apoptosis and the number of dying cells. Under these conditions, a subclone of MCF-7 cells resistant to 1,25-(OH)2D3 was isolated and characterized. In this early report, treatment of both parental and resistant MCF-7 cells with TAM was found to induce apoptosis and CLU expression (Welsh, 1994). The morphological changes typical of apoptotic death seen in 1,25-(OH)2D3-treated MCF-7 cells were also associated with upregulation of cathepsin B (a gene previously found associated with mammary gland apoptosis) and downregulation of bcl-2, an antiapoptotic gene. Thus, changes in the level of expression of several molecular markers show that the growth inhibitory effect of 1,25-(OH)2D3 on MCF-7 cells involves activation of apoptosis. A vitamin D3-resistant variant (MCF-7D3Res cells) was studied to gain further information in this model. The resistant clone was selected by continuous culture of MCF-7 cells in 100 nM 1,25-(OH)2D3. The experiment yielded the MCF-7D3Res cells, which represent a stably selected phenotype. In contrast to MCF-7 parental cells (MCF-7WT cells), MCF-7D3Res cells do not exhibit apoptotic morphology, DNA fragmentation, or upregulation of apoptosis-related proteins after treatment with 1,25-(OH)2D3. MCF-7D3Res cells are cross-resistant to several vitamin D3 analogues, all being potent growth regulators of MCF-7WT cells. Interestingly, apoptosis can be induced at comparable levels in both MCF-7WT and MCF-7D3Res cells in response to the antiestrogen TAM. Under these conditions, upregulation of CLU was seen in both cell lines (Narvaez *et al.*, 1996). Thus, cross-resistance to vitamin D3 analogues does not require CLU overexpression: on the contrary, even in resistant cells, induction of apoptosis by

another drug (i.e., TAM) is linked to *CLU* gene overexpression. This observation was then confirmed later in a study in which the role of apoptosis in
vitamin D-mediated growth arrest of MCF-7 cells was studied. The morphological features of apoptosis and the increase in the cell death rate were
accompanied by increased soluble DNA–histone complexes, indicative of
DNA fragmentation. Under these conditions, the level of CLU mRNA
and protein were significantly upregulated in MCF-7 cells treated with
1,25-(OH)2D3 as compared to control cells together with cathepsin B
mRNA (Simboli-Campbell *et al.*, 1996).

   The link between apoptosis induced by vitamin D3 analogues and CLU
overexpression was further investigated later on. A non hypercalcemic
vitamin D3 analogues called BXL-628 was developed to inhibit prostate
cell growth in benign prostate hyperplasia (BPH) and brought from the
bench to the animal model and finally to the clinics. In this study, *CLU*
gene overexpression was studied and used as a biomarker of cell response to
the analogue. In particular, CLU overexpression was specifically found
confined to cells positive for DNA fragmentation and cell death (Crescioli
et al., 2003, 2004).

XII. EARLY OBSERVATIONS: CLU EXPRESSION IS INDUCED BY TGF-BETA 1; NF-KAPPA B INHIBITORS BLOCKED THE SECRETION OF CLU

   Interestingly, a link between CLU expression and TGF-beta 1 was found in
MCF-7 cells treated with TAM. As described above, DNA fragmentation is
induced in MCF-7 cells by TAM as detected by terminal deoxynucleotidyl
transferase-mediated dUTP-biotin end labeling (TUNEL). Northern blot
hybridization studies showed an increase in the amounts of CLU and
TGF-beta 1 mRNAs in MCF-7 cells after treatment with TAM. The authors
of the study suggest that the induction of apoptosis by TAM in MCF-7 cells
may be mediated by the secretion of active TGF-beta (Chen *et al.*, 1996).
The first demonstration that TGF-beta 1 can induce intracellular localization and nuclear targeting of CLU appeared in the literature the same year
(1996) (Reddy *et al.*, 1996).

   The fact that CLU can be retained inside the cells following inhibition of
secretion was observed when inhibitors of nuclear factor-kappa B (NF-κB)
were given to glial cells. In this work, LPS was givn and glial cells became
activated. This maneuver induced a dose-dependent increase in CLU expression. It was known that NF-κB plays a pivotal role in glial activation. For this
reason, the activity of NF-κB was inhibited by using aspirin and MG-132. Both
treatments blocked basal CLU secretion as well as LPS-induced CLU secretion

(Saura *et al.*, 2003). MG-132 is well known as a potent inhibitor of the proteasome. In a very recent work, potent accumulation of nCLU in prostate epithelial cells following administration of MG-132 was confirmed. The likely explanation derives from the discovery that CLU is constantly cleared from prostate cancer cells by efficient polyubiquitination (Rizzi *et al.*, 2009). This finding bears also important consequences on the possible role of CLU and nCLU in tumorigenesis, but these issues will be further developed in this book. To this regard, and particularly considering the link between CLU and NF-κB activity, please see the two chapters dedicated to CLU and inflammation (see chapter "CLU a multifacet protein at the crossroad of inflammation and autoimmunity") and CLU and tumorigenesis (chapter "Regulation of CLU gene expression by oncogenes and epigenetic factors: Implication for tumourigenesis" Vol. 105). As a matter of fact, a different kind of link between CLU and NF-κB activity was found later. Inside the cell, CLU was found capable to bind and stabilize IkB, an important inhibitor of NF-κB. IkB is known to bind and block NF-κB in the cytosol. This event would inhibit nuclear translocation of NF-κB and its transactivating activity. Thus, in the intracellular compartment (cytosol), CLU may act as a potent inhibitor of NF-κB (Santilli *et al.*, 2003). In consideration of the fact that NF-κB is known to play a major role in inflammation and tumorigenesis, this discovery opens a new scenery for the possible role of CLU in cell transformation.

XIII. 1996: nCLU FINALLY APPEARS IN THE LITERATURE

The first experimental evidences that, in addition to the well-characterized sCLU form which preferentially acts in the extracellular compartment including blood and body fluids, a nuclear form of CLU (nCLU) also exist were provided in the year 1996. CLU, when retained inside the cells, may have completely different biological effects. In the pioneer work by Reddy *et al.* an intracellular form of the protein was found to be induced and accumulated in the nucleus of two epithelial cell lines (HepG2 and CCL64) in response to treatment with TGF-beta (Reddy *et al.*, 1996). The study provided the first *in vitro* demonstration that two active in-frame ATG sites can be recognized and drive the production of different CLU protein forms. Initiation from the ATG in exon 2 would encode for the secretory form of CLU (sCLU). The other in-frame ATG is located 33 amino acids downstream of the latter. When this ATG is active, the protein would lack the hydrophobic signal sequence. The truncated CLU protein resulting, with a lower molecular weight by Western blot analysis, is supposed to be not recognized by microsomes and therefore not glycosylated. Thus, we must hypothesize a completely different pathway of maturation. In this case, nCLU is retained inside the cell and it can be found both in the cytoplasm and in the nucleus. Targeting to the nucleus was supposed to be due to the presence of SV40-like nuclear localization signals (Reddy *et al.*, 1996).

In another report of the same year, CLU expression was studied in populations of Shionogi carcinoma cells subjected to multiple cycles of androgen withdrawal and replacement (intermittent androgen suppression). Interestingly, the parent tumor is negative to CLU expression: considering that prostate epithelial cells do express basal levels of CLU, this initial observation suggests that expression of CLU is initially downregulated in cancer cells. After castration, the *CLU* gene was found actively expressed in degenerating cells, comprising the vast majority of the tissue specimen. Thus, CLU overexpression was initially confirmed as associated to cell death. After the first and subsequent cycles of androgen withdrawal, CLU was also found expressed in nonregressing tumors. Immunohistochemistry analysis demonstrated that androgen-dependent tumor cells accumulated CLU both in cytoplasm and nuclei after each cycle of intermittent androgen suppression. The proportion of cells showing nuclear staining for CLU increased with cycles before castration from 3% to 30% of total (Akakura *et al.*, 1996). These data have been interpreted as demonstrating the protective role of CLU against cell death. In the light of the previous considerations and what is known today, these findings allow us to suggest that nuclear staining of androgen-dependent tumor cells under androgen-ablation condition is a demonstration of the opposite, that is, the proapoptotic role of nCLU instead. This was clearly demonstrated later. In the work by Akakura *et al.*, more intense nCLU staining was reported in recurrent androgen-independent cells. This finding was interpreted by the authors as an "anomalous nuclear localization of CLU" which may be instrumental to inhibit early events in the apoptotic process. In other words, according to the authors, sequestration of CLU into the nucleus would inhibit a cytolytic-protection activity of sCLU or cytoplasmic CLU. In their opinion, this event would help the generation of androgen-independent stem cells in an androgen-depleted environment. Actually, thanks to many independent reports, it is now well recognized that nuclear staining of CLU is a hallmark of cell death. Thus, nCLU-positive cells would have been those in which apoptotic death has been triggered. In the same system, the authors found constitutive expression of CLU in the cytoplasm of a lesser amount of cells, estimated to be around 2–3% of total. Cancer stem cells are supposed to be a minor population in a prostate cancer tissue specimen. As a matter of fact, their estimation of the relative proportion of androgen-independent stem cells along the intermittent androgen-suppression cycles showed that their number increased from 1/520,000 to 1/720, still less than 2% of total cells at final stage, a value fully compatible with the proportion of cells positive for constitutive CLU expression in the cytoplasm (2–3%) and not in the nucleus (up to 30%). As we will discuss later, very similar data, that is, increased expression of CLU in the cytoplasm accompanied by inhibition of nuclear targeting of CLU, was found in stable cell clones obtained from prostate cancer cells surviving to CLU overexpression. These cells exhibit slow proliferation activity, genetic instability, and resistance to cell death (Scaltriti *et al.*, 2004a,b). Therefore, the results by Akakura *et al.* can be interpreted in an alternative way: in their experimental system, stem cancer cells would be the minor population of cells with exclusive staining of the cytoplasm, in which very likely inhibition of nuclear targeting of CLU would provide prosurvival and acquisition of resistance to cell death.

XIV. A ROLE FOR nCLU IN THE NUCLEUS: BINDING TO Ku70

Few years later, CLU was identified as a Ku70-binding protein (KUB1-KUB4) using yeast two-hybrid analyses. Previously identified as KUB1, by sequence analysis it was then revealed to be CLU. CLU was previously described by the same research group as X-ray-inducible transcript 8 (XIP8) because it was found potently induced when high doses of X-ray were given to cells. KUB2 was identified as Ku80, while KUB3 and KUB4 were not identified in this work. Stable overexpression of full-length CLU in MCF-7:WS8 cell line caused decreased Ku70/Ku80 DNA end binding. Administration of CLU monoclonal antibodies rescued the cells (Yang *et al.*, 1999). This result is difficult to interpret and not in agreement with a prosurvival role of sCLU, because anti-CLU antibodies administered to living cells have very likely reacted with the secreted form of CLU and not the intracellular form, which was not accessible in the experimental system. In any case, a further confirmation of the proapoptotic role of nCLU appeared in the literature the year after (Yang *et al.*, 2000). In another work by the same research group, CLU was found coimmunoprecipitated and colocalized with Ku70/Ku80 by confocal microscopy *in vivo*. Ku70/Ku80 was defined as a DNA damage molecular sensor with a key role in repairing double-strand breaks in human MCF-7:WS8 breast cancer cells. Overexpression of nCLU or its minimal Ku70-binding domain (120 aa of CLU C-terminus) in nonirradiated MCF-7:WS8 cells dramatically reduced cell growth and colony-forming ability. Clonogenic toxicity was concomitant with cell-cycle arrest at the G_1 checkpoint and increased cell death. The authors concluded that enhanced expression and accumulation of nCLU–Ku70/Ku80 complexes is an important cell death signal (Yang *et al.*, 2000). Furthermore, increased nCLU was found to prevent

DNA-PK-mediated end joining. This event stimulated cell death in response
to IR. A further important confirmation appeared in the literature in 2001,
when overexpression of nCLU alone, in the absence of IR, was found to
stimulate cell death (Leskov *et al.*, 2001).

XV. nCLU TODAY: THE DARK SIDE OF THE MOON REVEALED

After 2003, several papers detecting nCLU and describing, or hypothesizing,
its possible biological role started to pile up in the literature. Our contribution to the field, besides the confirmation of the existence and the importance
of nCLU in other experimental systems, was the demonstration that overexpression of nCLU is associated to detachment-induced apoptosis, also
known as "anoikis" (Caccamo *et al.*, 2003, 2004, 2005). Following the
apoptotic stimulus (serum starvation), SV40-immortalized prostate epithelial cells (PNT1A) started to round up and slowly detach from plates. Then,
chromatin condensation and formation of apoptotic bodies were seen.
Under these conditions, caspase-3, -8, and -9 were activated together with
PARP, and DNA fragmentation was induced as seen by FACS analysis.
Anoikis-death was preceded by the appearance of CLU into the nucleus.
In our hands, a specific nCLU electrophoretic band of 45 kDa appeared as an
early event just before cell irreversible death induction. Detection of this
nCLU form to the nucleus was associated to S-phase arrest at the G0/G1-
S-phase checkpoint and incomplete DNA synthesis, consistently to what
we have previously found by transient overexpression of CLU cDNA in
the same cells (Bettuzzi *et al.*, 2002). Therefore, the proteomic expression
profile of CLU can change, at least in PNT1A cells, as a consequence of cell
growth condition. Nuclear targeting of a specific 45-kDa nCLU form was an
early event preceding cell death. This finding was consistent with the complementary discovery that *CLU* was identified as one of the rare genes whose
expression is specifically repressed by anchorage in a DNA microarray
analysis study (Goldberg *et al.*, 2001). The most important experimental
proof that a truncated form of CLU actually induces anoikis-death was
provided shortly after (Scaltriti *et al.*, 2004a,b). To this aim, we have
designed a family of expression vectors in which we have been deleted the
first in-frame ATG present on Isoform 2, but common to all alternative CLU
mRNAs and the leader peptide, from the full-length cDNA for CLU only
leaving the last in-frame ATG in the coding sequence (please see also chapter
"Clusterin (CLU): From one gene and two transcripts to many proteins" in
this book for further details on CLU transcripts isoforms). Therefore, these
vectors express a truncated form of CLU whose molecular weight is about

49 kDa as calculated by only considering the translated amino acid sequence and no further modifications such as glycosylation or cleavage. Consistently, we have actually detected a 49-kDa band in the experimental system, demonstrating that the lacking of the leader peptide actually completely abolished the known pathway of maturation, blocking the protein precursor from entering the Golgi apparatus for full processing and production of the secreted form. Under these conditions, the 49-kDa CLU form accumulated in the cytoplasm and in the nuclei (nCLU) and secretion of CLU was inhibited. Transient transfection of metastatic prostate cancer PC-3 cells with 49-kDa nCLU resulted in NLS-independent massive nuclear localization of the protein. G2/M-phase blockade and caspase-dependent apoptosis rapidly followed. Constitutive expression of 49-kDa nCLU in recombinant PC-3 cell clones caused intense clonogenic toxicity, confirming a negative correlation between nCLU accumulation and cell survival. Only rare clones overexpressing the truncated form of CLU lacking the leader peptide and targeting the nucleus survived, demonstrating that the primary role of nCLU is proapoptotic. The straightforward conclusion arising from these data is that expression of CLU protein and targeting to the nucleus is incompatible with the survival of human prostate cancer cells. Anyhow, when we engineered cells to maintain nCLU expressed, some clones survived. In these cells, we found that proliferation was inhibited and the phenotype was altered. In particular, mitosis was impaired, causing endo-reduplication and chromosome instability. Cells selected to survive nCLU overexpression were slow growing, due to arrest at the G2-M checkpoint by downregulation of the mitotic complex cyclin B1/cyclin-dependent kinase 1. The most intriguing feature of these cells is that, although stably transfected and expressing the 49-kDa nCLU form, the protein was exclusively localized in the cytoplasm. This scenario suggested that, to survive nCLU expression, prostate cancer cells had developed apoptosis-inhibitory properties including inhibition of nuclear targeting of CLU. This could have significance for the resistance of prostate cancers to chemo/radiotherapy, where CLU overexpression and cytoplasm localization of CLU has been observed, also opening new perspectives for the characterization of androgen-independent and apoptosis-resistant prostate cancer cells.

These experimental evidences may possibly help to explain the conflicting results obtained in different laboratories. CLU might be a proapoptotic or a survival gene depending from tissue contest, cell localization, and the prevailing protein form. The idea that sCLU may protect cells from death (to this regard, please see chapters "The chaperone action of CLU and its putative role in quality control of extracellular protein folding," "Cell protective functions of secretory CLU (sCLU)," this volume; and "CLU and chemoresistance," Vol. 105) while nCLU is a prodeath signal, is growing among the experts in the field. Further support to this hypothesis and to the idea that possible differential roles are played by the two CLU forms on cell growth and motility was recently provided (Moretti *et al.*, 2007). In nonmalignant PNT1A cells both sCLU and nCLU, when differentially overexpressed, significantly decreased cell proliferation and motility. In metastatic PC-3 cells instead, only nCLU inhibited cell growth and the migratory behavior, while sCLU was ineffective. This result strongly suggests a different role for CLU forms in benign and cancer cells, nicely explaining some contradictory data in the field. Inhibition of cell motility/ migration by nCLU was linked to a dramatic remodeling of the actin cytoskeleton. In particular, a massive dismantling of cytoskeleton organization was detected. This effect was due to specific binding of nCLU to α-actinin, a key protein for the regulation of actin cytoskeleton. Thus, a native protein partner of CLU was identified inside the cell. In addition, this finding nicely explains two important phenomena previously described (i) induction of anchorage-dependent apoptosis (anoikis-death) by nCLU and, conversely, (ii) repression of CLU expression by anchorage.

XVI. THE MOLECULAR MECHANISMS OF PRODUCTION OF nCLU ARE STILL SKETCHY

As described above, nuclear translocation of CLU was seen following ionizing radiation. The nCLU-inducible protein would bind Ku70 and triggers apoptosis. Also overexpression of CLU was found to kill MCF-7 cells. In this experimental model, an inactive precursor of nCLU was supposed to be constitutively expressed and reside in the cytoplasm when MCF-7 cells are not irradiated. X-ray administration would induce the translocation of CLU into the nucleus and induction of apoptosis would follow (Leskov *et al.*, 2003).

In this paper, the synthesis of endogenous nCLU is suggested to result by alternative splicing of the known CLU transcript. As a matter of fact, this event has never been confirmed later in other cell systems, leaving this possibility as only confined to the specific cell line used. On the other hand, the story became a lot more complicated recently when other transcripts coding for CLU were discovered. To this regard, please see also chapter "Clusterin (CLU): From one gene and two transcripts to many proteins" in this book.

An alternative explanation of how nCLU may be produced in the same system was provided by treating MCF-7 with TNF-α or the antiestrogen ICI 182,780, compounds both capable to induce apoptosis in MCF-7. In this experimental model, it has been described that the treatment caused substantial changes in the activity of Golgi-resident enzymes. As a consequence, the biogenesis of CLU was supposed to be significantly altered (O'Sullivan *et al.*, 2003).

Under these conditions, the appearance of a 50–53 kDa uncleaved, non-glycosylated, disulfide-linked form of CLU was described. This nCLU pro-death form was found accumulating in the nucleus. Surprisingly, in spite of the fact that CLU contains a cryptic SV-40-like NLS, mutation of this sequence did not affect nuclear targeting of nCLU form. More recently we found, by sequence comparison analysis, that three NLS are actually present in the full-length cDNA sequence of CLU. In our work, we succeeded at mutating all three of them, but again nuclear targeting of CLU resulted was unaffected (Scaltriti *et al.*, 2004b).

At the moment, the precise functional role of these cryptic NLS is not clear. O'Sullivan *et al.* suggested that the retrograde transport from the Golgi to the ER is responsible for the translocation of CLU to the nucleus. In this case, CLU would be not correctly processed and thus being nonglycosylated and uncleaved. Anyhow, they provided further evidences that translocation of CLU to the nucleus occurred in dying cells. As described before, we consistently found that removal of the leader sequence led to the production of a prodeath 49-kDa protein band in intact live cells, a molecular weight perfectly corresponding to the expected size of the nonglycosylated and uncleaved protein form (Scaltriti *et al.*, 2004b). These data do not support the hypothesis that translocation of CLU to the nucleus would occur as a late event due to disruption of nuclear membrane following the activation of cell death, but are suggesting instead that targeting of CLU to the nucleus is an early event occurring in living cells and triggering cell death later.

Again working on MCF-7 cells, the possible involvement of p53 at regulating CLU expression and cell response was hypothesized (Criswell *et al.*, 2003). Here sCLU, the extracellular secreted form of CLU, was found induced after very low, nontoxic doses of ionizing radiation and regarded as being cytoprotective. On the other hand, nCLU induced at higher X-ray dosage was confirmed cytotoxic. It appeared rationale to believe that suppression of sCLU (and maybe induction of nCLU) may stimulate cell death. In this contest, the factors possibly regulating the stress-inducible expression of sCLU have been studied. In this model, p53 was found to suppress sCLU induction responses. They used isogenically matched HCT116 colon cancer cell lines which differed only in p53 or p21 status. The experiments demonstrated a role for p53 in the transcriptional repression of sCLU. The finding is interesting, but similar experiments in the colon cancer model did not confirm a functional role of p53 at regulating the cell fate through CLU expression (Chen *et al.*, 2004). As a matter of fact, APC and p21 were found capable to induce CLU overexpression and nCLU translocation, triggering cell death. In the same paper is also shown that drug resistance by colon cancer cells can be reversed by upregulation of nCLU, which restored sensitivity to apoptosis in 5Fu + Fas treated resistant cells. Restored sensitivity was abolished by antisense against CLU (Chen et al., 2004). In this scenario, the idea that cells must presumably suppress sCLU to stimulate cell death is possible, but induction of nCLU seems to commit cells to death more potently. A recent finding supporting this idea is the fact that CLU was found to be a very short-lived protein (Rizzi et al., 2009). At difference from what was believed for long time, CLU protein forms are very unstable inside prostate cancer cells because they are constantly degraded by polyubiquitination and proteasome activity. Half-life of CLU forms is less than 2 h in prostate cancer cells. Interestingly, when protesome activity is specifically blocked with MG-132, CLU secretion is abolished and a huge accumulation of all CLU protein forms occurs inside the cells. Under these conditions, fluorescence microscopy demonstrated strong nuclear targeting of CLU (nCLU) and induction of cell death. This result suggests that prostate cancer cells may need to constantly deplete intracellular environment from nCLU and expel sCLU as secreted product to avoid commitment to death.

In 2005, a strong suggestion that production of nCLU may be a posttranslational event was provided by the discovery that intracellular calcium depletion by BAPTA-AM was capable to induce nuclear targeting of CLU with 4 h in prostate cells. Production of nCLU was again confirmed associated to apoptotic cell death induction (Caccamo et al., 2005). Due to the rapid targeting, induction of alternative transcription starting sites or alternative splicing seems unlikely. In this system, the prevailing form of CLU in the cells consists of a 65-kDa intracellular CLU precursor. This protein product is progressively converted to low-molecular-weight forms including sCLU. In cells grown in a calcium-depleted medium, the 65-kDa CLU precursor was rapidly converted to a 45-kDa CLU form. This event was preceding or concomitant to poly (ADP-ribose) polymerase (PARP) activation. Cells were rescued by Ca^{2+} administration, which blocked both the accumulation of the 45-kDa form and PARP activation. Administration of BAPTA-AM, a well-known intracellular Ca^{2+} blocker known to trigger caspases activation and apoptosis, caused a switch to the 45-kDa form of CLU already at 4 h after the beginning of treatment, a time when caspase-9 activation started to be evident. Only later, at 24–48 h, the decrease in cell proliferation was massive. Nuclear targeting of the 45-kDa CLU form was confirmed by fluorescence immunocytochemistry. Further insights on the biology of CLU and calcium are provided in the chapter "Regulation of CLU expression by calcium" of this book.

Rapid changes in the proteomic profile of CLU protein forms and subsequent accumulation in the cells was also seen in heat-shock experimental models (Caccamo et al., 2006). Sublethal heat shock, to which nontumorigenic PNT1A and metastatic PC-3 cells can cope and survive, inhibited the secretion of CLU (sCLU), leading to increased intracellular/cytoplasm accumulation of CLU. At difference, lethal heat stress caused massive accumulation of proapoptotic nCLU. Death was due to caspase-3-dependent apoptosis. To induce thermo-tolerance, double heat stress (sublethal heat shock followed by recovery and lethal stress) was given to both cell lines. The treatment induced HSP70 and thermo-tolerance in PNT1A cells, but not in PC-3 cells. Interestingly, CLU secretion was inhibited and CLU was accumulated in PNT1A, suggesting that targeting of CLU to the cytoplasm might be prosurvival. In PC-3 cells instead, accumulation of nCLU was very evident and concomitant to caspase-3 induction and PARP activation, that is, inducing apoptotic doom. Thus, CLU expression and its subcellular localization were strictly related to the cell fate. In particular, nCLU and HSP70 seems to affect cell survival in an antagonistic fashion. Prevalence of heat-induced nCLU would not allow metastatic PC-3 cells to cope with heat shock. An important clinical link of this finding is that this result may provide a rational explanation about why malignant cells are more sensitive to heat when delivered by minimally invasive procedures for ablation of localized prostate cancer. Of notice, in this experimental system the nuclear translocation of CLU was a very rapid event, not easy to explain by alternative site transcription or alternative splicing events. To further support this idea, we found that induction of nCLU was clearly associated to inhibition of CLU mRNA transcription, a data consistent with posttranslational modification of a protein precursor already present in the cell (Caccamo *et al.*, 2004). The precursor could be previously accumulated in the cell and kept "in-check," waiting for the apoptotic stimulus triggering its rapid conversion to the prodeath form.

XVII. IN THE END: IS TARGETING OF CLU TO THE NUCLEUS AN EXPERIMENTAL ARTIFACT?

One of the criticisms that can be raised on this issue is whether the experimental methods used to localize CLU inside the cell may have been misleading the researchers, leading to false conclusions. Disruption of the cell and purification of cell fractions is known to be tricky. Purity of the nuclear or cytoplasm fractions with no cross-contamination is always a problem. In addition, one of the best known features of CLU is its tendency to bind many proteins and cell structures once the cell is broken. CLU can bind to the nuclei of dead cells if lacking an intact cell membrane (Humphreys *et al.*, 1997). Homogenization of tissues and cells may release CLU forms which have not undergone complete processing and maturation. This may cause binding of CLU immature forms to cell structures, including nuclei. This event would not occur in intact cells under physiological conditions. For these reasons, the experimental procedure used to identify the location of CLU in intact cells is fundamental. The use of fluorescence or laser confocal microscopy in intact fixed cells may be of great help in this matter. As described before, a clear demonstration that CLU can actually be found in the nucleus of dying cells was provided by Scaltriti et al., thanks to these methods (Scaltriti et al., 2004a,b). In this model, an expression vector in which a truncated cDNA for human CLU lacking the leader peptide sequence (with its 5' leading ATG) was inserted. Under these conditions, a downstream 5' ATG in-frame became active, causing the production of a truncated form of CLU of 49 kDa which is perfectly compatible with an unglycosylated protein product. The accumulation of this CLU form in the cell nucleus caused cell death. This result was independently confirmed by Zhang et al. (2006) and Moretti et al. (2007), also showing that the truncated form of CLU was capable to specifically bind alpha-actinin, disturb the cytoskeleton organization, and cause cell detachment and inhibition of migration. Another interesting model in which CLU was found to enter the nucleus and cause rapid cell death was described by Caccamo et al. (2005). As previously described, depletion of intracellular calcium by administration of BAPTA-AM caused the translocation of CLU to the nucleus very rapidly. Only 4 h later the most of CLU was inside the nucleus and the vast majority of cells were subjected to anoikis-death.

But another important confirmation was obtained in vivo in a real clinical setting. Pucci et al. have studied the tissue distribution of CLU protein in healthy colon mucosa and cancer specimens by immunohistochemistry. This study has demonstrated that translocation of CLU from the nucleus (nCLU) to the cytoplasm is directly related to tumor progression. In fact, nCLU has been detected in the epithelial cells of healthy colon mucosa. Considering that we have previously seen that CLU is induced by APC in colon cells in a p21-dependent fashion (Chen et al., 2004) and that nCLU induces anoikis-death in epithelial cells (Caccamo et al., 2003), this finding is consistent with the proapoptotic action of nCLU and its role in the negative regulation of cell-cycle progression of colon cells. At difference, when colon cells progressed towards high-grade and metastatic carcinoma, nCLU was not detectable and CLU was distributed only in the cytoplasm of cancer cells (Pucci et al., 2004).

In all these experimental models, cells were not fractionated and CLU localization was detected in intact cells with appropriate experimental methods. Therefore, alternative explanations based on possible experimental artifacts are very unlikely to provide the answer.

XVIII. CONCLUSIONS

At the moment, it is still very true that unambiguous structural identification of alternative and intracellular forms of CLU are lacking. It is a personal and commonly shared experience that no known antibody (commercially available) is actually capable to discriminate sCLU from nCLU, no matter about which claims are made from the provider. At the same time, many good-quality antibodies can detect CLU when outside or inside the cell, thus allowing tracking extracellular, sCLU as well as intracellular CLU even when located inside the nucleus (nCLU) in different experimental models. The choice of the right tool for these studies is clearly fundamental. In any case, it is evident that more work is needed for a better understanding of this intriguing issue. Now the existence of different CLU forms with a differential propensity toward the cytosol and excretion in the extracellular compartment, or the nucleus and retention inside the cell, are firmly established. A general consensus toward the fact that a shift from sCLU to nCLU would doom the cell to death is now reached by experts. In many experimental systems, nCLU has been found to induce cell death. These notions may bear important consequences and practical implementation in the clinics soon.

Nevertheless, unravelling the alternative CLU protein structures, the mechanisms of their production, the differential regulation of their expression, and their potential mechanism of action through interaction with native intracellular partners or receptors will require more work in the future.

ACKNOWLEDGMENTS

Grant sponsor: FIL 2008 and FIL 2009, University of Parma, Italy; AICR (UK) Grant No. 06–711; Istituto Nazionale Biostrutture e Biosistemi (INBB), Roma, Italy.

REFERENCES

Agarwal, N., *et al.* (1996). Immunocytochemical colocalization of clusterin in apoptotic photoreceptor cells in retinal degeneration slow rds mutant mouse retinas. *Biochem. Biophys. Res. Commun.* **225**, 84–91.

Akakura, K., *et al.* (1996). Effects of intermittent androgen suppression on the stem cell composition and the expression of the TRPM-2 (clusterin) gene in the Shionogi carcinoma. *J Steroid Biochem. Mol. Biol.* **59**, 501–511.

Bettuzzi, S., *et al.* (1989). Identification of an androgen-repressed mRNA in rat ventral prostate as coding for sulphated glycoprotein 2 by cDNA cloning and sequence analysis. *Biochem. J.* **257**, 293–296.

Bettuzzi, S., *et al.* (1991). *In vivo* accumulation of sulfated glycoprotein 2 mRNA in rat thymocytes upon dexamethasone-induced cell death. *Biochem. Biophys. Res. Commun.* **175**, 810–815.

Bettuzzi, S., *et al.* (2002). Clusterin (SGP-2) transient overexpression decreases proliferation rate of SV40-immortalized human prostate epithelial cells by slowing down cell cycle progression. *Oncogene* **21**, 4328–4334.

Blaschuk, O., *et al.* (1983). Purification and characterization of a cell-aggregating factor (clusterin), the major glycoprotein in ram rete testis fluid. *J. Biol. Chem.* **258**, 7714–7720.

Caccamo, A. E., *et al.* (2003). Nuclear translocation of a clusterin isoform is associated with induction of anoikis in SV40-immortalized human prostate epithelial cells. *Ann. N. Y. Acad. Sci.* **1010**, 514–519.

Caccamo, A. E., *et al.* (2004). Cell detachment and apoptosis induction of immortalized human prostate epithelial cells are associated with early accumulation of a 45 kDa nuclear isoform of clusterin. *Biochem. J.* **382**, 157–168.

Caccamo, A. E., *et al.* (2005). Ca^{2+} depletion induces nuclear clusterin, a novel effector of apoptosis in immortalized human prostate cells. *Cell Death Differ.* **12**, 101–104.

Caccamo, A. E., *et al.* (2006). Nuclear clusterin accumulation during heat shock response: Implications for cell survival and thermo-tolerance induction in immortalized and prostate cancer cells. *J. Cell. Physiol.* **207**, 208–219.

Chen, H., *et al.* (1996). Tamoxifen induces TGF-beta 1 activity and apoptosis of human MCF-7 breast cancer cells *in vitro. J. Cell. Biochem.* **61**, 9–17.

Chen, T., *et al.* (2004). Clusterin-mediated apoptosis is regulated by adenomatous polyposis coli and is p21 dependent but p53 independent. *Cancer Res.* **64**, 7412–7419.

Cheng, C. Y., *et al.* (1988). Structural analysis of clusterin and its subunits in ram rete testis fluid. *Biochemistry* **27**, 4079–4088.

Crescioli, C., *et al.* (2003). Inhibition of spontaneous and androgen-induced prostate growth by a nonhypercalcemic calcitriol analog. *Endocrinology* **144**, 3046–3057.

Crescioli, C., *et al.* (2004). Inhibition of prostate cell growth by BXL-628, a calcitriol analogue selected for a phase II clinical trial in patients with benign prostate hyperplasia. *Eur. J. Endocrinol.* **150**, 591–603.

Criswell, T., *et al.* (2003). Repression of IR-inducible clusterin expression by the p53 tumor suppressor protein. *Cancer Biol. Ther.* **2**, 372–380.

Diemer, V., *et al.* (1992). Expression of porcine complement cytolysis inhibitor mRNA in cultured aortic smooth muscle cells. Changes during differentiation *in vitro. J. Biol. Chem.* **267**, 5257–5264.

Fritz, I. B., *et al.* (1983). Ram rete testis fluid contains a protein (clusterin) which influences cell–cell interactions *in vitro. Biol. Reprod.* **28**, 1173–1188.

Furlong, E. E., *et al.* (1996). Expression of a 74-kDa nuclear factor 1 (NF1) protein is induced in mouse mammary gland involution. Involution-enhanced occupation of a twin NF1 binding element in the testosterone-repressed prostate message-2/clusterin promoter. *J. Biol. Chem.* **271**, 29688–29697.

Gobe, G. C., *et al.* (1995). Clusterin expression and apoptosis in tissue remodeling associated with renal regeneration. *Kidney Int.* **47**, 411–420.

Goldberg, G. S., *et al.* (2001). Global effects of anchorage on gene expression during mammary carcinoma cell growth reveal role of tumor necrosis factor-related apoptosis-inducing ligand in anoikis. *Cancer Res.* **61**, 1334–1337.

Grassilli, E., *et al.* (1991). Studies on the relationship between cell proliferation and cell death: Opposite patterns of SGP-2 and ornithine decarboxylase mRNA accumulation in PHA-stimulated human lymphocytes. *Biochem. Biophys. Res. Commun.* **180**, 59–63.

Humphreys, D., *et al.* (1997). Effects of clusterin overexpression on TNFalpha- and TGFbeta-mediated death of L929 cells. *Biochemistry* **36**, 15233–15243.

Hurwitz, A., *et al.* (1996). Follicular atresia as an apoptotic process: Atresia-associated increase in the ovarian expression of the putative apoptotic marker sulfated glycoprotein-2. *J. Soc. Gynecol. Investig.* **3**, 199–208.

Jaggi, R., *et al.* (1996). Regulation of a physiological apoptosis: Mouse mammary involution. *J. Dairy Sci.* **79**, 1074–1084.

Jomary, C., *et al.* (1993). Comparison of clusterin gene expression in normal and dystrophic human retinas. *Brain Res. Mol. Brain Res.* **20**, 279–284.

Kane, R., *et al.* (2002). Transcription factor NFIC undergoes N-glycosylation during early mammary gland involution. *J. Biol. Chem.* 277, 25893–25903.

Laping, N. J., *et al.* (1994). Transforming growth factor-beta 1 induces neuronal and astrocyte genes: Tubulin alpha 1, glial fibrillary acidic protein and clusterin. *Neuroscience* 58, 563–572.

Leskov, K. S., *et al.* (2001). When X-ray-inducible proteins meet DNA double strand break repair. *Semin. Radiat. Oncol.* 11, 352–372.

Leskov, K. S., *et al.* (2003). Synthesis and functional analyses of nuclear clusterin, a cell death protein. *J. Biol. Chem.* 278, 11590–11600.

Masamune, A., *et al.* (2002). Ligands of peroxisome proliferator-activated receptor-gamma induce apoptosis in AR42J cells. *Pancreas* 24, 130–138.

Min, B. H., *et al.* (2003). Clusterin expression in the early process of pancreas regeneration in the pancreatectomized rat. *J. Histochem. Cytochem.* 51, 1355–1365.

Montpetit, M. L., *et al.* (1986). Androgen-repressed messages in the rat ventral prostate. *Prostate* 8, 25–36.

Moretti, R. M., *et al.* (2007). Clusterin isoforms differentially affect growth and motility of prostate cells: Possible implications in prostate tumorigenesis. *Cancer Res.* 67, 10325–10333.

Moulson, C. L., and Millis, A. J. (1999). Clusterin (Apo J) regulates vascular smooth muscle cell differentiation *in vitro. J. Cell. Physiol.* 180, 355–364.

Narvaez, C. J., *et al.* (1996). Characterization of a vitamin D3-resistant MCF-7 cell line. *Endocrinology* 137, 400–409.

Norman, D. J., *et al.* (1995). The lurcher gene induces apoptotic death in cerebellar Purkinje cells. *Development* 121, 1183–1193.

O'Sullivan, J., *et al.* (2003). Alterations in the post-translational modification and intracellular trafficking of clusterin in MCF-7 cells during apoptosis. *Cell Death Differ.* 10, 914–927.

Pampfer, S., *et al.* (1997). Increased cell death in rat blastocysts exposed to maternal diabetes *in utero* and to high glucose or tumor necrosis factor-alpha *in vitro. Development* 124, 4827–4836.

Pucci, S., *et al.* (2004). Modulation of different clusterin isoforms in human colon tumorigenesis. *Oncogene* 23, 2298–2304.

Ranganna, K., *et al.* (2003). Gene expression profile of butyrate-inhibited vascular smooth muscle cell proliferation. *Mol. Cell. Biochem.* 254, 21–36.

Reddy, K. B., *et al.* (1996). Transforming growth factor beta (TGF beta)-induced nuclear localization of apolipoprotein J/clusterin in epithelial cells. *Biochemistry* 35, 6157–6163.

Rizzi, F., *et al.* (2009). Clusterin is a short half-life, poly-ubiquitinated protein, which controls the fate of prostate cancer cells. *J. Cell. Physiol.* 219, 314–323.

Santilli, G., *et al.* (2003). Essential requirement of apolipoprotein J (clusterin) signaling for IkappaB expression and regulation of NF-kappaB activity. *J. Biol. Chem.* 278, 38214–38219.

Saura, J., *et al.* (2003). Microglial apolipoprotein E and astroglial apolipoprotein J expression *in vitro*: Opposite effects of lipopolysaccharide. *J. Neurochem.* 85, 1455–1467.

Scaltriti, M., *et al.* (2004a). Clusterin overexpression in both malignant and nonmalignant prostate epithelial cells induces cell cycle arrest and apoptosis. *Br. J. Cancer* 91, 1842–1850.

Scaltriti, M., *et al.* (2004b). Intracellular clusterin induces G2-M phase arrest and cell death in PC-3 prostate cancer cells1. *Cancer Res.* 64, 6174–6182.

Simboli-Campbell, M., *et al.* (1996). 1,25-Dihydroxyvitamin D3 induces morphological and biochemical markers of apoptosis in MCF-7 breast cancer cells. *J. Steroid Biochem. Mol. Biol.* 58, 367–376.

Sivamurthy, N., *et al.* (2001). Apolipoprotein J inhibits the migration, adhesion, and proliferation of vascular smooth muscle cells. *J. Vasc. Surg.* 34, 716–723.

Thomas-Salgar, S., and Millis, A. J. (1994). Clusterin expression in differentiating smooth muscle cells. *J. Biol. Chem.* 269, 17879–17885.

Welsh, J. (1994). Induction of apoptosis in breast cancer cells in response to vitamin D and antiestrogens. *Biochem. Cell Biol.* **72**, 537–545.

Witzgall, R., *et al.* (1994). Localization of proliferating cell nuclear antigen, vimentin, c-Fos, and clusterin in the postischemic kidney. Evidence for a heterogenous genetic response among nephron segments, and a large pool of mitotically active and dedifferentiated cells. *J. Clin. Invest.* **93**, 2175–2188.

Yang, C. R., *et al.* (1999). Isolation of Ku70-binding proteins (KUBs). *Nucleic Acids Res.* **27**, 2165–2174.

Yang, C. R., *et al.* (2000). Nuclear clusterin/XIP8, an x-ray-induced Ku70-binding protein that signals cell death. *Proc. Natl. Acad. Sci. USA* **97**, 5907–5912.

Zhang, X., *et al.* (1997). Internucleosomal DNA fragmentation is not obligatory for castration induced rat ventral prostate cell apoptosis *in vivo*. *Cell Death Differ.* **4**, 304–310.

Zhang, Q., *et al.* (2006). The leader sequence triggers and enhances several functions of clusterin and is instrumental in the progression of human prostate cancer *in vivo* and *in vitro*. *BJU Int.* **98**, 452–460.

The Chaperone Action of Clusterin and Its Putative Role in Quality Control of Extracellular Protein Folding

Amy Wyatt, Justin Yerbury, Stephen Poon, Rebecca Dabbs, and Mark Wilson

School of Biological Sciences, University of Wollongong, Wollongong, New South Wales 2522, Australia

The function(s) of clusterin may depend upon its topological location. A variety of intracellular "isoforms" of clusterin have been reported but further work is required to better define their identity. The secreted form of clusterin has a potent ability to inhibit both amorphous and amyloid protein aggregation. In the case of amorphous protein aggregation, clusterin forms stable, soluble high-molecular-weight complexes with misfolded client proteins. Clusterin expression is increased during many types of physiological and pathological stresses and is thought to function as an extracellular chaperone (EC). The pathology of a variety of serious human diseases is thought to arise as a consequence of the inappropriate aggregation of specific extracellular proteins (e.g., Aβ peptide in Alzheimer's disease and β_2-microglobulin

Advances in CANCER RESEARCH
Copyright 2009, Elsevier Inc. All rights reserved.

0065-230X/09 $35.00
DOI: 10.1016/S0065-230X(09)04006-8

in dialysis-related amyloidosis). We have proposed that together with other abundant ECs (e.g., haptoglobin and α_2-macroglobulin), clusterin forms part of a previously unknown quality-control (QC) system for protein folding that mediates the recognition and disposal of extracellular misfolded proteins via receptor-mediated endocytosis and lysosomal degradation. Characterizing the mechanisms of this extracellular QC system will thus have major implications for our understanding of diseases of this type and may eventually lead to the development of new therapies. © 2009 Elsevier Inc.

I. INTRODUCTION

The focus of this chapter is on the chaperone action of the secreted extracellular form of clusterin. However, before dealing with this, a brief discussion of what is currently known about intra- and extracellular forms of CLU follows. While in mammals clusterin is usually secreted, many reports have suggested that it occurs in the nucleocytosol continuum, where it has been proposed to influence, for example, DNA repair (Yang *et al.*, 2000), transcription (Santilli *et al.*, 2003), microtubule organization (Kang *et al.*, 2005), and apoptosis (Debure *et al.*, 2003; Yang *et al.*, 2000; Zhang *et al.*, 2005). Various mechanisms have been proposed to explain the presence of clusterin in the nuclear and cytosolic compartments, including:

(i) Alternative initiation of transcription—this was proposed to yield a 43-kDa nuclear clusterin isoform in CCL64 and HepG2 (two epithelial cell lines) treated with transforming growth factor beta (Reddy *et al.*, 1996).

(ii) Alternative splicing—this was proposed to account for a 49-kDa nuclear clusterin isoform in MCF-7 cells exposed to ionizing radiation (Leskov *et al.*, 2003).

(iii) Retrotranslocation of clusterin via an ERAD-like pathway from the Golgi apparatus to the cytosol (Nizard *et al.*, 2007).

(iv) Reinternalization of secreted clusterin from the extracellular milieu into the cytosol (Kang *et al.*, 2005).

However, it is important to note that none of these studies sequenced the intracellular isoforms; therefore, it is not known for certain whether clusterin isoforms of different mass resulted from alternative splicing or transcription initiation or simply represented species at different stages of maturation (e.g., cleaved or uncleaved, at different stages of glycosylation). Furthermore, in some cases, the methods used to identify the location of intracellular isoforms (e.g., cell fractionation) may have led to false conclusions. It is known that clusterin binds to the nuclei of dead cells lacking an intact cell membrane (Humphreys *et al.*, 1997). Disruption of live unfixed cells may release poorly glycosylated and/or uncleaved clusterin from the ER/Golgi allowing it to bind to nuclei, which would not normally occur in intact cells. In addition, a very recent report describes the secreted form of clusterin as being present in the nucleocytosolic continuum and mitochondria (Trougakos *et al.*, 2009). Unambiguous structural identification of intracellular isoforms of clusterin protein (e.g., "nuclear clusterin") is required before their existence can be accepted as firmly established and their function(s) meaningfully assigned. A recent study reported that a 60-kDa intracellular clusterin isoform conferred resistance to apoptosis upon cancer cells by antagonizing conformationally altered Bax (Zhang *et al.*, 2005); however, this study did not address how clusterin was diverted from its normal secretory pathway.

Determining the biological importance of clusterin has been complicated by the propensity of the protein to interact with a wide variety of molecules including lipids (de Silva *et al.*, 1990), amyloid proteins (Boggs *et al.*, 1996), components of the complement membrane attack complex (MAC) (Jenne and Tschopp, 1989; Kirszbaum *et al.*, 1989), immunoglobulins (Wilson and Easterbrook-Smith, 1992), other molecules of clusterin (Blaschuk *et al.*, 1983), misfolded proteins (Wilson and Easterbrook-Smith, 2000), and the chemotherapeutic drug Taxol (Park *et al.*, 2008). A reasonable hypothesis is that many of these interactions result from a single underlying property of clusterin, related to a primary function. There is now a substantial body of evidence that clusterin has a chaperone action like that of the small heat-shock proteins (sHSPs) and is one of the first described extracellular chaperones (ECs) (Carver *et al.*, 2003; Hatters *et al.*, 2002; Humphreys *et al.*, 1999; Lakins *et al.*, 2002; Poon *et al.*, 2000, 2002a,b; Wilson and Easterbrook-Smith, 2000; Yerbury *et al.*, 2007). As part of its chaperone action, clusterin appears to bind to exposed regions of hydrophobicity on misfolded proteins (Poon *et al.*, 2002a). As many of the identified ligands for clusterin are hydrophobic in character, we suggest that the interactions of clusterin with these ligands are related to its chaperone properties. In the case of the clusterin–Taxol interaction, this may have clinical consequences; sequestration of Taxol by clusterin may account for why clusterin-overexpressing tumors are Taxol-resistant (Park *et al.*, 2008). However, this is not to exclude the possibility of other functions for clusterin. Indeed, the function(s) of the protein may be dependent upon its location (intracellular vs. extracellular). This is one of the reasons why more careful work is needed to definitively establish whether any of the variously reported cytosolic/nuclear clusterin isoforms are genuine. The rest of this chapter will focus on what we believe is a major function for the protein in the context of extracellular space.

II. CLUSTERIN AS A CHAPERONE

A. Molecular Chaperones

Increased expression in response to cellular stresses, including heat shock, oxidative stress, and heavy metals, or to pathologic conditions, such as infection, inflammation, ischemia, tissue damage, and mutant proteins associated with genetic diseases, is synonymous with, but not limited to, a group of stress-response proteins broadly referred to as molecular chaperones. The term "molecular chaperone" was first used to describe the functional activity of the nuclear protein, nucleoplasmin, in preventing the aggregation of folded histone proteins with DNA during nucleosomal assembly. This term was later extended by Ellis to encompass a large and diverse group of proteins that facilitate the noncovalent assembly of other proteins but which do not themselves form part of the final folded structure (Ellis, 1987). Much of the current understanding of molecular chaperones centers on the heat-shock protein (HSP) family, a large group of structurally unrelated thermal-stress-inducible intracellular proteins, which includes HSP100, HSP90, HSP70, HSP60, and the sHSPs. These molecular chaperones function to: (1) ensure correct folding of newly synthesized polypeptides, (2) unfold and refold polypeptides during translocation across membranes, (3) assemble and disassemble macromolecular complexes, or (4) resolubilize or facilitate the degradation of partially folded and/or aggregated proteins.

From a therapeutic perspective, by far the most important trait of molecular chaperones is their ability to specifically recognize, interact with, and prevent the aggregation of nonnative proteins. Protein aggregation is thought to underpin the pathologies of many debilitating human diseases, known as protein deposition diseases, such as cataract, cystic fibrosis, and the amyloidoses (e.g., Alzheimer's, Parkinson's, and Creutzfeldt–Jakob diseases). Whether *in vivo* or *in vitro*, protein aggregation occurs when a specific trigger induces the partial unfolding of proteins, which causes once buried hydrophobic regions to become surface-exposed. Due to their aversion to the aqueous surrounding, the exposed hydrophobic regions of nonnative proteins associate with similar regions on neighboring proteins. Continuation of this molecular cascade leads to the formation of protein aggregates which usually adopt one of two major forms. They may exist as disorganized amorphous species comprised of randomly associated proteins, or as highly ordered fibrillar structures. For example, it has been proposed that the cytotoxic species associated with various neurodegenerative disorders are most likely to be soluble oligomeric protein aggregates that are amorphous in structure, while the protein aggregates found in the deposits of amyloidoses sufferers are characterized by highly ordered β-sheet-rich proteins in the form of insoluble fibrils (Dobson, 2003, 2004; Lopez de la Paz and Serrano, 2004). It is believed that molecular chaperones such as the HSPs interact with hydrophobic domains of nonnative "client" proteins. Through this mechanism, molecular chaperones are able to promiscuously bind and stabilize nonnative proteins, but avoid natively structured proteins.

B. The *In Vitro* Chaperone Action of Clusterin

A decade ago, a report emerged that provided, for the first time, detailed evidence to support the hypothesis that clusterin may function as a molecular chaperone (Humphreys et al., 1999). A series of publications soon followed, which, together with the original article, highlighted clusterin's ability to inhibit, in a dose-dependent manner, the stress-induced amorphous aggregation and precipitation of a number of unrelated target proteins, including heat-stressed glutathione-S-transferase, catalase, alcohol dehydrogenase, and ovotransferrin, and chemically reduced bovine serum albumin, α-lactalbumin, lysozyme, and ovotransferrin (Humphreys et al., 1997; Poon et al., 2000, 2002a,b). Enzyme-linked immunosorbent assays (ELISAs) and size-exclusion chromatography indicated that, like the sHSPs, clusterin binds preferentially to and forms high-molecular-weight complexes with these stressed target proteins. It was also shown that clusterin interacts only with slowly precipitating forms of γ-crystallin and lysozyme, indicating that the kinetics of protein aggregation may be an important determinant for efficient binding of clusterin to its client proteins (Poon et al., 2002b).

The manner in which clusterin interacts with partially unfolded and aggregating client proteins is not clear. Sequence analysis of full-length clusterin reveals several motifs that may be important for its chaperone action. Amphipathic α-helices located at the N-terminus of the α-subunit and at both the N- and C-termini of the β-subunit possess hydrophobic surfaces that are believed to mediate clusterin binding to cell membranes of sperm (Law and Griswold, 1994) and potentially, to various other ligands, many of which are predominantly hydrophobic in character. It is likely that the binding of clusterin to many of its ligands is achieved through these hydrophobic moieties. Results published by Poon et al. (2002a) are consistent with this proposed mechanism. At physiological pH, clusterin exists predominantly as a mixture of dimers and tetramers of the disulfide-linked α–β heterodimer "monomer." When exposed to mildly acidic conditions, these oligomers dissociate to form α–β monomers which show enhanced binding to the hydrophobic probe 1-anilino-8-naphthalenesulfonate (ANS), indicating significant exposure of hydrophobic surfaces. The exposed hydrophobic regions appear to be available for binding to chaperone client proteins, because the mildly acidic conditions induce increased client protein binding by clusterin and an enhancement of its chaperone action (Poon *et al.*, 2002a). In a physiological context, this effect may be of significant importance. At sites of inflammation, cardiac ischemia, infarcted brain, and in the brains of Alzheimer's sufferers (McGeer and McGeer, 1997), a phenomenon known as acidosis occurs, where the local pH can fall below 6. Under these mildly acidic conditions, the enhancement of clusterin's chaperone action could help reduce the rate of progression or severity of pathology by inhibiting the aggregation and accumulation of inflammatory and/or toxic insoluble protein deposits.

Despite its ability to stabilize aggregation-prone proteins, clusterin is unable to protect against or restore the stress-induced loss of enzyme activity of alcohol dehydrogenase, catalase, and glutathione-*S*-transferase. The lack of recovery of functional activities of these enzymes implies that clusterin is unable to independently refold them. Nevertheless, it was shown that clusterin can maintain stressed proteins in a state that is competent for subsequent refolding by heat-shock cognate protein 70 (HSC70) (Poon *et al.*, 2000). Several sHSPs have been shown to act in a similar manner to clusterin. For instance, HSP18.1 (from peas) was shown to stabilize heat-denatured firefly luciferase in such a way that other chaperones, present at high concentrations in rabbit reticulocytes or wheat germ extracts, could refold it (Lee *et al.*, 1997). Similarly, mammalian HSP25 was shown to form stable complexes with heat-stressed citrate synthase, which could be subsequently reactivated by HSP70-mediated refolding (Ehrnsperger *et al.*, 1997).

Earlier reports had described the ability of clusterin to inhibit the *in vitro* fibrillar aggregation of the amyloid β (Aβ) peptide, the major component of senile plaques associated with Alzheimer's disease (Matsubara *et al.*, 1995; Oda *et al.*, 1995). Subsequent studies showed that clusterin, at a concentration commonly present in human plasma (100 μg/mL), can reduce by as much as 90% the fibrillar aggregation of PrP106–126, the amyloid-forming and neurotoxic fragment of the prion protein responsible for the bovine prion disease, bovine spongiform encephalopathy (BSE) (McHattie and Edington, 1999). In 2002, Hatters and coworkers demonstrated similar inhibitory effects of clusterin on amyloid fibril formation. It was reported that even at substoichiometric concentrations, equivalent to a 30:1 molar excess of client protein to clusterin, clusterin was able to completely suppress *in vitro* amyloid fibrillogenesis of apolipoprotein C-II (ApoC-II), a protein lipase activator (Hatters *et al.*, 2002). However, in this case, clusterin was unable to dissociate the mature ApoC-II fibrils once they had formed (Hatters *et al.*, 2002). More recently, clusterin has been shown to influence the fibrillogenesis of other proteins, including the disease-associated variants of lysozyme, I56T and D67H (Kumita *et al.*, 2007), α-synuclein, calcitonin, κ-casein, SH3 (amyloidogenic domain of phosphatidylinositol 3-kinase), and CCβw, a derivative of the coiled-coil β (CCβ) peptide, which was originally designed *de novo* as a model to study the transformation of helical conformations into amyloid fibrils (Yerbury *et al.*, 2007). For most of the targets tested, clusterin caused dose-dependent inhibition of amyloid fibril formation. However, for α-synuclein, Aβ, and calcitonin, very low levels of clusterin (relative to client protein) significantly increased amyloid formation. It was proposed that when present at very low concentrations, clusterin may stabilize prefibrillar oligomers that "seed" fibril growth and are believed to be primarily responsible for amyloid-associated cytotoxicity. In fact, it was shown that under these conditions the cytotoxicity of Aβ and SH3 aggregates were enhanced (Yerbury *et al.*, 2007). These results suggest that the clusterin:client protein ratio is an important determinant of the nature and extent of effects of clusterin on both amyloid formation and toxicity. This may partly explain why clusterin can exert significantly different effects on amyloid formation in different experimental systems. In all cases where inhibition was observed, electron microscopy imaging of the resulting aggregates revealed that the final structures formed in the presence of clusterin ranged from spherical particles of varying diameters to amorphous aggregates.

Clusterin does not bind to native forms of proteins, and affects amyloid formation to a greater extent when added during the lag phase versus the growth phase (Kumita *et al.*, 2007; Yerbury *et al.*, 2007). Exactly how clusterin interacts with amyloid-forming proteins remains unclear, although the existing evidence suggests that clusterin interacts predominantly with prefibrillar oligomeric species formed during the latter stages of the lag or nucleation phase of amyloid aggregation (Kumita *et al.*, 2007; Yerbury *et al.*, 2007). Prefibrillar aggregates are known to possess surface-exposed hydrophobicity that has been implicated in their ability to exert cytotoxic effects by interacting with cell membranes (Bucciantini *et al.*, 2004). Thus the interaction of clusterin with amyloid-forming proteins may, as in the case of amorphously aggregating proteins, stem from hydrophobic interactions. Another possibility is that clusterin may interact with amyloid-forming proteins via functional groups of charged or aromatic amino acids that are critical for fibril formation (Makin *et al.*, 2005). Regardless of the precise mechanism by which clusterin affects amyloid formation, it is possible that increasing the levels of clusterin *in vivo* may represent a potential therapeutic strategy in the fight against extracellular amyloid deposition disorders.

III. *IN VIVO* INSIGHTS INTO THE CHAPERONE ACTION OF CLUSTERIN

A. Associations of Clusterin with Extracellular Protein Deposits

Clusterin is found in human blood plasma at around 35–105 μg/ml (Murphy *et al.*, 1988), in cerebrospinal fluid (CSF) at 1.2–3.6 μg/ml (Choi *et al.*, 1990), in seminal plasma at around 2–15 mg/ml (Choi *et al.*, 1990), and is also found in numerous other biological fluids including breast milk, ocular fluid, and urine (Aronow *et al.*, 1993). Due to the inherent chaperone action of clusterin, and its presence in many extracellular fluids, it follows that it may be found in extracellular locations associated with misfolded proteins. In fact, clusterin has been found associated with all disease-associated extracellular protein deposits tested to date, including amyloid deposits (Yerbury *et al.*, 2007) (Table I). The presence of clusterin in these deposits may indicate its incorporation into insoluble aggregates under conditions in which the capacity of extracellular protein folding quality-control (QC) mechanisms have been exceeded.

The overexpression of clusterin has been reported in many disease states including atherosclerosis, cancers, diabetes, myocardial infarction, neurodegenerative diseases, and renal diseases (Rosenberg and Silkensen, 1995). Additionally, clusterin is upregulated in experimental models of pathological

Table I Protein Deposition Disorders in Which Clusterin Has Been Found Colocalized with Extracellular Protein Deposits

Disease	Main constituent	Reference
Alzheimer's disease	Aβ	Calero *et al.* (2000)
Creutzfeldt–Jakob disease	PrP	Freixes *et al.* (2004)
Gerstmann–Straussler–Scheinker disease	PrP	Chisea *et al.* (1996)
Gelatinous drop-like corneal dystrophy	Keratoepithelin	Nishida *et al.* (1999)
Lattice type I corneal dystrophy	M1S1	Nishida *et al.* (1999)
Age-related macular degeneration	Drusen	Crabb *et al.* (2002), Sakaguchi *et al.* (1998)
Pseudoexfoliation (PEX) syndrome	PEX material	Zenkel *et al.* (2006)
Down's syndrome	Aβ	Kida *et al.* (1995)
HCHWA-Dutch type	Aβ	Matt-Schieman *et al.* (1996)
Familial British dementia	ABri	Ghiso *et al.* (1995)
Atherosclerosis	LDL/ApoB100	Witte *et al.* (1993)

stress including heat stress (Michel *et al.*, 1997), heavy metal exposure (Trougakos *et al.*, 2006), inhibition of the proteosome generating proteotoxic stress (Loison *et al.*, 2006), ionizing radiation (Criswell *et al.*, 2005), oxidative stress (Strocchi *et al.*, 2006), and shear stress (Ubrich *et al.*, 2000). This further suggests that clusterin plays a role in stress states associated with many types of disease. It may also act in a protective capacity in pseudoexfoliation (PEX) syndrome, as decreased levels of clusterin in the eye are associated with the deposition of PEX material causing this disease (Zenkel *et al.*, 2006). In contrast, it has been proposed that at least under some conditions, clusterin can promote the solubility and toxicity of oligomers of Aβ peptide (Oda *et al.*, 1995). A later related study claimed that clusterin had the opposite effect, and inhibited the toxicity of similar Aβ oligomers (Boggs *et al.*, 1996). A recent study has provided an observation that may at least partly explain these apparently contradictory results—the molar ratio of clusterin to fibril-forming protein determines whether pro- or antiamyloidogenic effects result (Yerbury *et al.*, 2007).

B. Mouse Models

There have been a number of studies undertaken using transgenic mice in which clusterin expression has been ablated ("knock out" mice) or (in one case only) increased. The results from these studies have been apparently contradictory, suggesting that simple interpretations based on the assumption that whatever effect is observed results from the action of clusterin alone are not appropriate. In a transgenic mouse model of Alzheimer's disease using mice expressing amyloid precursor protein (APP; PDAPP mice), a comparison between *clusterin*$^{-/-}$ and *clusterin*$^{+/+}$ mice showed no difference in Aβ deposition in the brain at 12 months of age. However, the ablation of clusterin expression resulted in lower thioflavin-S staining of brain matter and a decrease in the number of visibly damaged neurons (DeMattos *et al.*, 2002). Although there was no electron micrograph data to confirm that these deposits were in fact amyloid the authors suggested that clusterin promoted fibril formation and associated toxicity in this model. In contrast, PDAPP double knockout *ApoE*$^{-/-}$, *clusterin*$^{-/-}$ mice showed earlier onset and had significantly increased Aβ levels in cerebrospinal and interstitial fluid and thioflavin-S-positive material in the brain (DeMattos *et al.*, 2004), suggesting a cooperative protective effect of ApoE and clusterin to inhibit Aβ deposition in the brain. Collectively, these results suggest that clusterin's *in vivo* effects on amyloid formation are likely to involve multiple interactions and form part of a complex mechanism that is currently not fully understood. To better understand the nature and mechanism(s) of interaction between clusterin and amyloid-forming proteins in the complex *in vivo* environment it is essential that further *in vitro* studies are undertaken.

One study has suggested that clusterin can promote cell injury—Han *et al.* (2001) reported that clusterin knockout mice had 50% less brain injury than control mice following neonatal hypoxic-ischemic injury. In contrast, a number of other studies suggest that clusterin has protective effects:

- In an axotomy-induced cell death model, $clusterin^{+/+}$ mice had significantly more surviving motor neurons compared to $clusterin^{-/-}$ mice, suggesting a neuroprotective role (Wicher and Aldskogius, 2005).
- In clusterin knockout mice, damage to testicular cells is increased after heat shock and the clearance of damaged cells is impaired (Bailey *et al.*, 2002).
- In mice overexpressing clusterin, the number of dead cells present post-ischemic brain injury was significantly lower than those found in wild-type and $clusterin^{-/-}$ mice. Also, in this model, the content of glial and inflammatory cells in the area surrounding the ischemic zone was least in clusterin overexpressing mice and greatest in clusterin knockout mice (wild-type mice were in between these two values; (Wehrli *et al.*, 2001)).
- $Clusterin^{-/-}$ mice have more severe inflammation and cellular pathology in experimentally induced murine autoimmune myocarditis than clusterin-expressing control mice (McLaughlin *et al.*, 2000).
- Lastly, ageing $clusterin^{-/-}$ mice develop progressive glomerulopathy which is characterized by the accumulation of insoluble protein deposits in the kidneys (Rosenberg *et al.*, 2002). This suggests that clusterin inhibits the age-dependent accumulation of protein deposits in glomeruli.

Taken together, the mouse model studies suggest that clusterin (i) can affect extracellular protein aggregation, and (ii) may be involved in the clearance of aggregating protein species and/or dead cells which could have important anti-inflammatory and cytoprotective consequences *in vivo*. The report from Han *et al.* (2001) proposing that clusterin exacerbates cell injury may relate to a unique function of clusterin in that model, and/or may imply that the inherent complexity of the system is inconsistent with the assumption that the effects seen arise solely from the absence of clusterin.

IV. OTHER EXTRACELLULAR CHAPERONES (ECs) AND A MODEL FOR QUALITY CONTROL (QC) OF EXTRACELLULAR PROTEIN FOLDING

Denatured plasma proteins are catabolized more rapidly than their native counterparts (Margineanu and Ghetie, 1981) and it has been shown that polymorphonuclear leukocytes selectively catabolize denatured extracellular proteins versus those that are in native conformations (Bocci *et al.*, 1968). In both cases, lysosomal enzymes are implicated in the degradation process (Bouma, 1982; Coffey and de Duve, 1968). In addition, the fact that each plasma protein has an individual and specific half-life suggests that they are selectively removed from the extracellular space and degraded (Margineanu and Ghetie, 1981). Furthermore, there is a positive correlation between pathological protein deposition and ageing; extracellular protein aggregation and deposition is usually only observed in disease states and in the elderly (Carrell, 2005). Extracellular pathological protein deposits include Aβ in Alzheimer's disease, human islet amyloid polypeptide in Type II diabetes, and prion protein in Creutzfeldt–Jakob disease. Collectively, these observations suggest that there is an extracellular protein folding QC system that under normal conditions keeps extracellular protein aggregation in check but that loses its efficiency with age (Dobson, 2002). If this virtually uncharacterized extracellular protein folding QC system is analogous to the extensively characterized intracellular system (Wickner *et al.*, 1999), then it is likely that it incorporates multiple functional components. The discovery of clusterin as an EC may be just the tip of an iceberg in terms of characterizing mechanisms of extracellular protein folding QC.

A. Other ECs

In recent years a number of secreted proteins have been reported as having an *in vitro* chaperone action. These include serum amyloid P component (Coker *et al.*, 2000), fibrinogen (Tang *et al.*, 2009), albumin (Marini *et al.*, 2005), macrophage inhibitory factor (Cherepkova *et al.*, 2006), apolipoprotein E (Wood *et al.*, 1996), haptoglobin (Yerbury *et al.*, 2005a), and α_2-macroglobulin (French *et al.*, 2008). Not all studies have been thoroughly followed up and it is not clear in all cases how the *in vitro* chaperone activity relates to any corresponding *in vivo* function. Apart from clusterin, the best characterized ECs are haptoglobin and α_2-macroglobulin.

B. Haptoglobin

Haptoglobin is a secreted glycoprotein found in most body fluids. Its plasma concentration is between 0.3 and 2 mg/mL (Bowman and Kurosky, 1982) and it is found in CSF between 0.5 and 2 μg/mL (Sobek and Adam, 2003). Humans may exhibit one of three different haptoglobin phenotypes (Hp 1–1, Hp 1–2, or Hp 2–2). In its simplest form, it exists in a barbell-like structure of two disulphide linked $\alpha_1\beta$ subunits (Hp 1–1); in the presence of the α_2 chain haptoglobin assembles into a range of various sized disulphide linked polymers of $\alpha\beta$ subunits in a $\alpha_1\beta-(\alpha_2\beta)-\alpha_1\beta$ configuration (Hp 2–1), or in $\alpha_2\beta$-based ring-like structures (Hp 2–2) (Kurosky *et al.*, 1980). Haptoglobin is an acute-phase protein, the expression of which is upregulated in response to a variety of insults including infection, neoplasia, pregnancy, trauma, acute myocardial infarction, and other inflammatory conditions (Langlois and Delanghe, 1996). Haptoglobin has many described biological functions, however, it is most commonly known for its particularly strong noncovalent interaction with hemoglobin ($K_D \sim 10^{-15}$ M) (Dobryszycka, 1997). The sequestration of hemoglobin by haptoglobin reduces oxidative damage after induced hemolysis (Lim *et al.*, 1998) through the reduction of free hemoglobin and iron available to catalyse oxidation reactions (Gutteridge, 1987). Haptoglobin has also been described as having a bacteriostatic effect (Barclay, 1985), an ability to stimulate angiogenesis through an unknown mechanism (Cid *et al.*, 1993), and has been implicated in the regulation of lymphocyte transformation (Baskies *et al.*, 1980).

Recently, haptoglobin has been characterized as one of a small group of secreted proteins that can act as molecular chaperones by inhibiting the stress-induced aggregation of a range of proteins (Ettrich *et al.*, 2002; Pavlicek and Ettrich, 1999; Yerbury *et al.*, 2005a, 2009). Functionally, it has a sHSP-like ability to maintain aggregating proteins in solution by forming high-molecular-weight complexes with them. Moreover, depletion of haptoglobin from human plasma significantly increases the susceptibility of plasma proteins to aggregation and precipitation *in vitro* (Yerbury *et al.*, 2005a).

C. α_2-Macroglobulin

α_2-Macroglobulin is a secreted glycoprotein that is highly abundant in many biological fluids. Its approximate concentrations in human plasma and CSF are 1.5–2 mg/mL and 1.0–3.6 μg/mL, respectively (Biringer *et al.*, 2006; Sottrup-Jensen, 1989). α_2-Macroglobulin is a homotetramer consisting of 180-kDa subunits that form two disulfide-linked homodimers that noncovalently interact to form the 720-kDa tetramer (Jensen and Sottrup-Jensen, 1986). α_2-Macroglobulin is best known for its ability to bind to and inhibit a wide range of proteases. It does this using a unique trapping mechanism. α_2-Macroglobulin contains a "bait region" which when proteolytically cleaved exposes a thioester bond which results in covalent attachment to the protease and a large conformational change in the α_2-macroglobulin tetramer exposing a recognition site for the low-density lipoprotein receptor-related protein (LRP), a cell-surface receptor (Barrett and Starkey, 1973). Aside from its known interaction with proteases, α_2-macroglobulin is known to bind to many growth factors and cytokines as well as being found complexed with proteins known to aggregate and form deposits *in vivo* such as β_2-microglobulin (Gouin-Charnet *et al.*, 2000), Aβ peptide (Du *et al.*, 1997), and prion protein (Adler and Kryukov, 2007).

Recent evidence indicates that α_2-macroglobulin is likely to be an important EC. Similar to clusterin and haptoglobin, α_2-macroglobulin has also been shown to have a shSP-like chaperone activity (French *et al.*, 2008) – it inhibits the *in vitro* formation of both amorphous and fibrillar aggregates by a range of proteins (French *et al.*, 2008; Yerbury *et al.*, 2009). This activity has been demonstrated for globular proteins and peptides not only at physiological temperature and pH, but also at elevated temperatures, low pH, and under conditions of oxidative stress. It has been shown that α_2-macroglobulin binds to and forms complexes with some intermediate protein species during the amyloid aggregation process (Yerbury *et al.*, 2009) and that depletion of α_2-macroglobulin from human plasma renders plasma proteins more susceptible to aggregation and precipitation at physiological temperatures (French *et al.*, 2008). While in complex with nonnative proteins, α_2-macroglobulin can still be activated by a protease and bind to LRP on the surface of cells (French *et al.*, 2008). After LRP-mediated endocytosis, complexes formed between α_2-macroglobulin and proteases are known to transit to lysosomes where they are degraded (Lauer *et al.*, 2001). Interestingly, complexes formed between α_2-macroglobulin and peptides elicit an immune response similar to that elicited by HSP–peptide complexes (Strivastava, 2002).

D. A Family of Functionally Related ECs

Clusterin, haptoglobin, and α_2-macroglobulin form a small group of secreted proteins with functional similarities. They have, in common, an ability to stably bind misfolded proteins and thereby inhibit inappropriate protein–protein interactions, preventing aggregation, and maintaining proteins in solution. Each of these ECs has specific cell-surface receptors that recognize and internalize them in complex with ligands for subsequent degradation in lysosomes (Graversen *et al.*, 2002; Hammad *et al.*, 1997; Shibata *et al.*, 2000). Each of these ECs is also found associated with pathological protein deposits *in vivo*. Clusterin is found associated with

extracellular protein deposits in numerous diseases including normal peripheral drusen and macular drusen in age-related macular degeneration patients
(Crabb *et al.*, 2002), MAC in renal immunoglobulin deposits (French *et al.*,
1992), prion deposits in Creutzfeldt–Jakob disease (Chisea *et al.*, 1996;
Freixes *et al.*, 2004; Sasaki *et al.*, 2002), and amyloid plaques and soluble
Aβ peptide in Alzheimer's disease (Calero *et al.*, 2000; Ghiso *et al.*, 1993)
(see also Table I). Haptoglobin has been found associated with amyloid
deposits in Alzheimer's disease (Powers *et al.*, 1981) and drusen (Kliffen
et al., 1995). In addition, α_2-macroglobulin is found associated with Aβ
deposits in Alzheimer's disease (Fabrizi *et al.*, 2001) and prion protein
deposits in spongiform encephalopathies (Adler and Kryukov, 2007). The
reason for the *in vivo* association of ECs with protein deposits is not yet
established, but it may indicate a failure in or the overwhelming of the
machinery responsible for QC of extracellular protein folding.

E. A Model for QC of Extracellular Protein Folding

We have proposed a model for the QC of extracellular protein folding in
which ECs recognize and bind to regions of exposed hydrophobicity on
nonnative extracellular proteins to form soluble complexes. These complexes are subsequently internalized via receptor-mediated endocytosis and
degraded by proteolytic enzymes within lysosomes (Fig. 1). The liver and
the reticuloendothelial system are likely to be the major contributors to the
clearance of EC–client protein complexes *in vivo* (Yerbury *et al.*, 2005b). If
this model is correct, one would expect to detect EC–client protein
complexes *in vivo*. In fact, both clusterin and α_2-macroglobulin are found
complexed with soluble amyloid-forming proteins in human extracellular
fluids (clusterin–Aβ (Ghiso *et al.*, 1993); α_2-macroglobulin–prion
protein (Adler and Kryukov, 2007); α_2-macroglobulin–β_2-microglobulin
(Motomiya *et al.*, 2003); clusterin–prion protein (Ecroyd *et al.*, 2005)).
In addition, it would be expected that inhibiting these *in vivo* clearance
mechanisms could induce extracellular protein deposition disorders such as
Alzheimer's disease. Evidence from studies in mice suggests that this could
be the case; radiolabeled Aβ injected in to the brain of mice is normally
rapidly removed, however, coinjection with the LDL family inhibitor
RAP, antibodies against LRP-1 or α_2-macroglobulin significantly slows
its removal (Shibata *et al.*, 2000). Furthermore, when complexed with
clusterin, Aβ_{1-42} is cleared from the brain more than 80% faster than
Aβ_{1-42} alone, and this clearance is significantly inhibited by antimegalin
antibodies (Bell *et al.*, 2007).

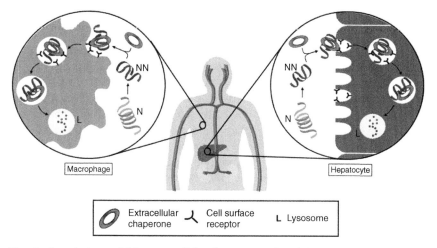

Fig. 1 Speculative model for extracellular chaperone-mediated clearance of nonnative proteins. Nonnative extracellular proteins are bound by extracellular chaperones that mediate their uptake into cells by receptor-mediated endocytosis. EC–NN complexes are internalized and moved by vesicular transport to lysosomes, where they are degraded. Receptors are recycled back to the cell surface (not shown). The primary sites of action of this process are likely to be the liver and the reticuloendothelial system. See key for other details. (Reproduced with permission from Yerbury *et al.*, 2005b). (See Page 6 in Color Section at the back of the book.)

V. FUTURE RESEARCH DIRECTIONS

A. Refinement of *In Vitro* Conditions Used to Study the Chaperone Action of Clusterin

The induction of protein aggregation *in vitro* requires conditions (such as increased temperature) which partially disrupt native structure but which do not completely inhibit interactions such as hydrogen bonding. However, the intrinsic stability of globular mammalian proteins means that, at 37 °C, low concentrations of most proteins in simple buffers do not unfold and aggregate at an experimentally convenient rate. There are obvious practical limitations to producing clusterin–client protein complexes *in vitro* to mimic those generated *in vivo*. If we are to consider extracellular proteins in an *in vivo* context then it is important to remember that additional stresses including ROS and shear stress will be present and contribute to the unfolding of these proteins. Moreover, macromolecular crowding such as that present in complex biological fluids would favor protein aggregation *in vivo* compared to buffered solutions at low protein concentrations. *In vitro*, it is difficult to accurately replicate the *in vivo* effects of, for example, shear stress, especially when it is experimentally inconvenient to wait for prolonged incubations spanning many days to weeks. Therefore, slightly higher than normal body temperatures are generally employed to induce protein unfolding for chaperone studies. While protein unfolding occurs in the same manner regardless of temperature (Day *et al.*, 2002), when studying the action of chaperones it is important to investigate them in a context that is as physiologically relevant as possible. Therefore, further investigations of the chaperone activity of clusterin and the formation of clusterin–client protein complexes should also include the development of *in vitro* systems to induce protein unfolding using conditions that are closer to those expected *in vivo*. In particular, the combined use of shear stress (equivalent to that possible under high arterial pressure 10–70 dyne/cm^2), macromolecular crowding (to reflect the crowding effect of extracellular biological fluids containing up to 70 mg/mL protein), and heat (less than or equal to 41 °C, the highest temperature for which human survival is expected) may be required to develop such systems *in vitro*.

B. Identification of Endogenous Chaperone Client Proteins

The identification of endogenous clusterin client proteins in biological fluids will be an important step toward establishing the role of clusterin in extracellular protein folding QC. Two approaches worth exploring are:

(1) Use clusterin as "bait" for client proteins after imposing a physiologically relevant stress in a biological fluid (e.g., mild heat and shear stress in plasma). While it is expected that clusterin-stressed protein complexes would be cleared very quickly *in vivo*, in the absence of mechanisms to dispose of them *in vitro*, proteins copurifying with clusterin after exposure to physiologically comparable stress (but not under "normal" conditions) would be strong candidates as endogenous clients. These putative client proteins could be identified by separating them by 2D-SDS polyacrylamide gel electrophoresis and subsequent analysis by mass spectrometry.

(2) Identify putative clusterin–client protein complexes in plasma by using ELISAs to capture clusterin and probe for bound client proteins using specific antibodies.

Identification of endogenous client proteins that are also implicated in protein deposition diseases (PDDs) would be of particular interest and would inform the potential development of clusterin in PDD-targeted therapies.

C. Clusterin–Client Protein Complexes

Given that clusterin appears to recognize stressed client proteins via exposed regions of hydrophobicity, a characteristic shared by all misfolded proteins, it is feasible and even likely that complexes formed between clusterin and misfolded proteins *in vivo* will contain more than one client protein. Moreover, clusterin may work together with the two other known ECs (α_2-macroglobulin and haptoglobin) *in vivo* to stabilize misfolded proteins, thus forming a diverse array of heterogeneous chaperone–client protein complexes under conditions of stress. Ideally, clusterin–client protein complexes could be purified directly from biological fluids, however, it is likely that after they are formed *in vivo* they are quickly taken up by tissues and degraded. Furthermore, traditional immunoaffinity methods require harsh denaturing conditions to elute the bound protein, which are likely to disrupt clusterin–client protein complexes. Therefore, the *in vitro* use of purified proteins is likely to be required for the bulk production and purification of clusterin–client protein complexes to be used in, for example, investigations of their *in vivo* fate in animal models. If, as predicted they are rapidly cleared from extracellular fluids such as plasma, the identification of key organs of uptake will be the first step toward elucidating the molecular clearance mechanisms involved. Furthermore, to provide further insights into the chaperone action of clusterin, the stoichiometry and structure of these complexes should be characterized. Physical characteristics such as size, relative exposed hydrophobicity, and surface charge may help to identify which specific endocytic receptors target these complexes *in vivo*.

D. Receptors

A number of investigations have focused on identifying possible interactions between members of the LDL receptor superfamily and clusterin (Bajari *et al.*, 2003; Bartl *et al.*, 2001; Calero *et al.*, 1999; Hammad *et al.*, 1997; Kounnas *et al.*, 1995; Lakins *et al.*, 2002; Mahon *et al.*, 1999; Zlokovic *et al.*, 1996). Cellular internalization of clusterin via the LDL receptor megalin was the first reported clusterin–LDL receptor superfamily interaction (Kounnas *et al.*, 1995). Subsequent reports described the internalization of free clusterin and clusterin–Aβ peptide complexes by the same receptor (Hammad *et al.*, 1997; Zlokovic *et al.*, 1996). Recently, two other human members of the LDL receptor superfamily, ApoE receptor 2 (ApoER2) and very low-density lipoprotein receptor (VLDLR), were reported to bind and internalize free clusterin and leptin–clusterin complexes using transfected cell models (Bajari *et al.*, 2003). Interactions of clusterin with chicken oocyte-specific LDL receptors have also been described (Mahon *et al.*, 1999).

A recent study has suggested that megalin and LRP are capable of mediating the clusterin-dependent clearance of cellular debris into nonprofessional phagocytes (Bartl *et al.*, 2001). However, the previous report of Kounnas *et al.* (1995) indicated that megalin, but not LRP, binds clusterin. Additional unidentified mechanisms of clusterin-dependent internalization were also suggested by Bartl *et al.* (2001). The affinity of binding of clusterin to megalin is increased by the association of clusterin with lipids (Calero *et al.*, 1999). It is currently unknown how binding interactions with other molecules, such as misfolded proteins, affect the binding affinity of clusterin for megalin or other members of the LDL receptor superfamily. The finding that clusterin has independent binding sites for megalin, misfolded proteins, and native ligands is consistent with a model in which misfolded extracellular proteins are cleared via clusterin-dependent receptor-mediated endocytosis (Lakins *et al.*, 2002). However, because the reported expression of megalin is limited to cells of the kidney, epididymis, lung, breast, thyroid, and eye (Lundgren *et al.*, 1997), other more abundant receptors may be involved in clearing clusterin–client protein complexes *in vivo*.

The identification of specific receptors involved in the uptake of clusterin–client protein complexes will be important to define some of the major elements of the model proposed for extracellular protein folding QC (Fig. 1). Potential approaches include using protein microarrays and/or surface plasmon resonance to screen candidate receptors (including LDL superfamily members and scavenger receptors), and transfected cell lines expressing specific receptors for their interaction with clusterin–client protein complexes. These complexes might bind to receptors via clusterin or possibly via site(s) on the misfolded client proteins. Although there are no known interactions between clusterin and scavenger receptors, considering that many scavenger receptors have the unique ability to recognize damaged or modified ligands and not their native counter parts, they are worthy of examination in this context. One possibility is that a shared physical characteristic of misfolded proteins held in clusterin–client protein complexes, such as disrupted secondary structure or increased exposed hydrophobicity, targets them to receptors for subsequent internalization and lysosomal degradation. However, it is also possible that clusterin–client protein complexes bind to receptors via a specific site(s) on clusterin, such as that proposed for megalin (Lakins *et al.*, 2002). Clearly, the mechanism responsible for mediating receptor binding may be receptor-specific and will have to be determined on a case-by-case basis once relevant receptor(s) are identified.

VI. CONCLUSION

Clusterin may exert different functions depending on its locale. There is much work to be done to better define why and how clusterin is sometimes found in various locations within cells. This chapter has focussed on what appears to be a major function for the protein in the extracellular context. A large number of serious human diseases are characterized by the deposition of extracellular protein aggregates. Our newly acquired knowledge of ECs presents an exciting avenue for the development of strategies to combat these diseases. However, these strategies are unlikely to be as simple as the upregulation of EC expression in affected individuals. Upregulation of clusterin is known to be associated with cancer progression and protects cells from chemotherapy drugs such as Taxol (Park *et al.*, 2008). Additionally, increasing the levels of α_2-macroglobulin, without increasing the throughput of the pathways that are known to clear the EC, has been reported to have detrimental effects (Fabrizi *et al.*, 2001). Therefore, it appears that ECs are a single player in a system for extracellular protein folding QC that is likely to involve other important elements such as endocytic cell-surface receptors. Characterization of the route by which clusterin–client protein complexes are disposed of *in vivo* will be critical to define the role of clusterin in extracellular protein folding QC and to underpin any attempts to develop new clusterin-based therapies for treating extracellular PDDs. While we have focussed here on the role of clusterin, it is predicted that all three known ECs (and possibly other yet to be identified ECs) act in similar ways to protect extracellular spaces from potentially pathological protein deposition. An important challenge is to identify how.

REFERENCES

Adler, V., and Kryukov, V. (2007). Serum macroglobulin induces prion protein transition. *Neurochem. J.* **1**, 43–52.

Aronow, B. J., Lund, S. D., Brown, T. L., Harmony, J. A. K., and Witte, D. P. (1993). Apolipoprotein J expression at fluid–tissue interfaces: Potential role in barrier cytoprotection. *Proc. Natl. Acad. Sci. USA* **90**, 725–729.

Bailey, R. W., Aronow, B., Harmony, J. A. K., and Griswold, M. D. (2002). Heat shock-initiated apoptosis is accelerated and removal of damaged cells is delayed in the testis of clusterin/apoJ knock-out mice. *Biol. Reprod.* **66**, 1042.

Bajari, T. M., Strasser, V., Nimpf, J., and Schneider, W. J. (2003). A model for modulation of leptin activity by association with clusterin. *FASEB J.* **17**, 1505–1507.

Barclay, R. (1985). The role of iron in infection. *Med. Lab. Sci.* **42**, 166–177.

Barrett, A. J., and Starkey, P. M. (1973). The interaction of alpha 2-macroglobulin with proteinases. Characteristics and specificity of the reaction, and a hypothesis concerning its molecular mechanism. *Biochem. J.* **133**, 709–724.

Bartl, M. M., Luckenbach, T., Bergner, O., Ulrich, O., and Koch-Brandt, C. (2001). Multiple receptors mediate apoJ-dependent clearance of cellular debris into nonprofessional phagocytes. *Exp. Cell Res.* **271,** 130–141.

Baskies, A. M., Chretien, P. B., Weiss, J. F., Makuch, R. W., Beveridge, R. A., Catalona, W. J., and Spiegel, H. E. (1980). Serum glycoproteins in cancer patients: First reports of correlations with *in vitro* and *in vivo* parameters of cellular immunity. *Cancer* **45,** 3050–3060.

Bell, R. D., Sagare, A. P., Friedman, A. E., Bedi, G. S., Holtzman, D. M., Deane, R., and Zlokovic, B. V. (2007). Transport pathways for clearance of human Alzheimer's amyloid beta-peptide and apolipoproteins E and J in the mouse central nervous system. *J. Cereb. Blood Flow Metab.* **27,** 909–918.

Biringer, R. G., Amato, H., Harrington, M. G., Fonteh, A. N., Riggins, J. N., and Huhmer, A. F. (2006). Enhanced sequence coverage of proteins in human cerebrospinal fluid using multiple enzymatic digestion and linear ion trap LC-MS/MS. *Breif. Funct. Genomic. Proteomic.* **5,** 144–153.

Blaschuk, O., Burdzy, K., and Fritz, I. B. (1983). Purification and characterization of a cell-aggregating factor (clusterin), the major glycoprotein in ram rete testis fluid. *J. Biol. Chem.* **258,** 7714–7720.

Bocci, V., Masti, L., Pacini, A., and Viti, A. (1968). Catabolism of native and denatured serum 131-I-proteins by polymorphonuclear leucocytes. *Exp. Cell Res.* **52,** 129–139.

Boggs, L. N., Fuson, K. S., Baez, M., Churgay, L., McClure, D., Becker, G., and May, P. C. (1996). Clusterin (apoJ) protects against *in vitro* amyloid beta (1–40) neurotoxicity. *J. Neurochem.* **67,** 1324–1327.

Bouma, J. M. W. (1982). Some aspects of plasma protein metabolism as compared with intracellular protein breakdown. *Acta Biol. Med. Ger.* **41,** 53–60.

Bowman, B. H., and Kurosky, A. (1982). Haptoglobin: The evolutionary product of duplication, unequal crossing over, and point mutation. *Adv. Hum. Genet.* **12,** 189–261.

Bucciantini, M., Calloni, G., Chiti, F., Formigli, L., Nosi, D., Dobson, C. M., and Stefani, M. (2004). Pre-fibrillar amyloid protein aggregates share common features of cytotoxicity. *J. Biol. Chem.* **279,** 31374–31382.

Calero, M., Tokuda, T., Rostagno, A., Kumar, A., Zlokovic, B., Frangione, B., and Ghiso, J. (1999). Functional and structural properties of lipid-associated apolipoprotein J (clusterin). *Biochem. J.* **344,** 375–383.

Calero, M., Rostagno, A., Matsubara, E., Zlokovic, B., Frangione, B., and Ghiso, J. (2000). Apolipoprotein J (clusterin) and Alzheimer's disease. *Microsc. Res. Tech.* **50,** 305–315.

Carrell, R. W. (2005). Cell toxicity and conformational disease. *Trends Cell Biol.* **15,** 574–580.

Carver, J. A., Rekas, A., Thorn, B. C., and Wilson, M. R. (2003). Small heat-shock proteins and clusterin: Intra- and extracellular molecular chaperones with a common mechanism of action and function. *IUBMB Life* **55,** 661–668.

Cherepkova, O. A., Lyutoba, E. M., Eronina, T. B., and Gurvits, B. Y. (2006). Chaperone-like activity of macrophage migration inhibitory factor. *Int. J. Biochem. Cell Biol.* **38,** 43–55.

Chisea, R., Angeretti, N., Lucca, E., Salmona, M., Tagliavini, F., Bugiani, O., and Forloni, G. (1996). Clusterin (SGP-2) induction in rat astroglial cells exposed to prion protein fragment 106–126. *Eur. J. Neurosci.* **8,** 589–597.

Choi, N. H., Tobe, T., Hara, K., Yoshida, H., and Tomita, M. (1990). Sandwich ELISA for quantitative measurement of SP40, 40 in seminal plasma and serum. *J. Immunol. Med.* **131,** 159–163.

Cid, M. C., Grant, D. S., Hoffman, G. S., Auerbach, R., Fauci, A. S., and Kleinman, H. K. (1993). Identification of haptoglobin as an angiogenic factor in sera from patients with systemic vascularitis. *J. Clin. Invest.* **91,** 977–985.

Coffey, J. W., and de Duve, C. (1968). Digestive activity of lysosomes. *J. Biol. Chem.* **243,** 3355.

Coker, A. R., Puris, A., Baker, D., Pepys, M. B., and Wood, S. P. (2000). Molecular chaperone properties of serum amyloid P component. *FEBS Lett.* **473**, 199–202.

Crabb, J. W., Miyagi, M., Gu, X., Shadrach, K., West, K. A., Sakaguchi, H., Kamei, M., Hasan, A., Yan, L., Rayborn, M. E., Salomon, R. G., and Hollyfield, J. G. (2002). Drusen proteome analysis: An approach to the etiology of age-related macular degeneration. *Proc. Natl. Acad. Sci. USA* **99**, 14682–14687.

Criswell, T., Beman, M., Araki, S., Leskov, K., Cataldo, E., Mayo, L. D., and Boothman, D. A. (2005). Delayed activation of insulin-like growth factor-1 receptor/Src/MAPK/Egr-1 signaling regulates clusterin expression, a pro-survival factor. *J. Biol. Chem.* **280**, 14212–14221.

Day, R., Bennion, B. J., Ham, S., and Daggett, V. (2002). Increasing temperature accelerates protein unfolding without changing the pathway of unfolding. *J. Biol. Chem.* **322**, 189–203.

Debure, L., Vayssiere, J. L., Rincheval, V., Loison, F., Le Drean, Y., and Michel, D. (2003). Intracellular clusterin causes juxtanuclear aggregate formation and mitochondrial alteration. *J. Cell Sci.* **116**, 3109–3121.

DeMattos, R. B., O'Dell, M. A., Parsadanian, M., Taylor, J. W., Harmony, J. A. K., Bales, K. R., Paul, S. M., Aronow, B., and Holtzman, D. M. (2002). Clusterin promotes amyloid plaque formation and is critical for neuritic toxicity in a mouse model of Alzheimer's disease. *Proc. Natl. Acad. Sci. USA* **99**, 10843–10848.

DeMattos, R. B., Cirrito, J. R., Parsadanian, M., May, P. C., O'Dell, M. A., Taylor, J. W., Harmony, J. A. K., Arnow, B. J., Bales, K. R., Paul, S. M., and Holtzman, D. M. (2004). ApoE and clusterin cooperatively suppress Abeta levels and deposition: Evidence that ApoE regulates extracellular Abeta metabolism *in vivo*. *Neuron* **41**, 193–202.

de Silva, II. V., Harmony, J. A. K., Stuart, W. D., Gil, C. M., and Robbins, J. (1990). Apolipoprotein J: Structure and tissue distribution. *Biochemistry* **29**, 5380–5389.

Dobryszycka, W. (1997). Biological functions of haptoglobin—New pieces to an old puzzle. *Eur. J. Clin. Chem. Clin. Biochem.* **35**, 647–654.

Dobson, C. M. (2002). Getting out of shape. *Nature* **418**, 729–730.

Dobson, C. M. (2003). Protein folding and disease: A view from the first Horizon Symposium. *Nat. Rev. Drug Discov.* **2**, 154–160.

Dobson, C. M. (2004). Principles of protein folding, misfolding and aggregation. *Semin. Cell Dev. Biol.* **15**, 3–16.

Du, Y., Ni, B., Glinn, M., Dodel, R. C., Bales, K. R., Zhang, Z., Hyslop, P. A., and Paul, S. M. (1997). Alpha2-macroglobulin as a beta-amyloid peptide-binding plasma protein. *J. Neurochem.* **69**, 299–305.

Ecroyd, H., Belghazi, M., Dacheux, J. L., and Gatti, J. L. (2005). The epididymal soluble prion protien forms a high-molecular-mass complex in association with hydrophobic proteins. *Biochem. J.* **392**, 211–219.

Ehrnsperger, M., Buchner, J., and Gaestel, M. (1997). Structure and Function of Small Heat Shock Proteins Marcel Dekker, New York.

Ellis, R. J. (1987). Proteins as molecular chaperones. *Nature* **328**, 378–379.

Ettrich, R., Brandt, W., Kopecky, V., Baumruk, V., Hofbauerova, K., and Pavlicek, Z. (2002). Study of chaperone-like activity of human haptoglobin: Conformational changes under heat shock conditions and localization of interaction sites. *Biol. Chem.* **383**, 1667–1676.

Fabrizi, C., Businaro, R., Lauro, G. M., and Fumagalli, L. (2001). Role of alpha2-macroglobulin in regulating amyloid-protein neurotoxicity: Protective or detrimental factor? *J. Neurochem.* **78**, 406–412.

Freixes, M., Puig, B., Rodriguez, A., Torrejon-Escribano, B., Blanco, R., and Ferrer, I. (2004). Clusterin solubility and aggregation in Creutzfeldt–Jakob disease. *Acta Neuropathol.* **108**, 295–301.

French, L. E., Tschopp, J., and Schifferli, J. A. (1992). Clusterin in renal tissue: Preferential localization with the terminal complement complex and immunoglobulin deposits in glomeruli. *Clin. Exp. Immunol.* **88**, 389–393.

French, K., Yerbury, J. J., and Wilson, M. R. (2008). Protease activation of alpha2-macroglobulin modulates a chaperone-like broad specificity. *Biochemistry* **47**, 1176–1185.

Ghiso, J., Matsubara, E., Koudinov, A., Choi-Miura, N. H., Tomita, M., Wisniewski, T., and Frangione, B. (1993). The cerebrospinal-fluid soluble form of Alzheimer's amyloid beta is complexed to SP-40,40 (apolipoprotein J), an inhibitor of the complement membrane-attack complex. *Biochem. J.* **293**, 20–30.

Ghiso, J., Plant, G. T., Revesz, T., Wisniewski, T., and Frangione, B. (1995). Familial cerebral amyloid angiopathy (British type) with nonneuritic amyloid plaque formation may be due to a novel amyloid protein. *J. Neurol. Sci.* **129**, 74–75.

Gouin-Charnet, A., Laune, D., Granier, C., Mani, J. C., Pau, B., Mourad, G., and Argiles, A. (2000). Alpha2-macroglobulin, the main serum antiprotease, binds beta2-microglobulin, the light chain of the class I major histocompatibility complex, which is involved in human disease. *Clin. Sci.* **98**, 427–433.

Graversen, J. H., Madsen, M., and Moestrup, S. K. (2002). CD163: A signal receptor scavenging haptoglobin–hemoglobin complexes from plasma. *Int. J. Biochem. Cell Biol.* **34**, 309–314.

Gutteridge, J. M. (1987). The antioxidant activity of haptoglobin towards haemoglobin-stimulated lipid peroxidation. *Biochim. Biophys. Acta* **917**, 219–223.

Hammad, S. M., Ranganathan, S., Loukinova, E. B., Twal, W. O., and Argraves, W. S. (1997). Interaction of apolipoprotein J-amyloid B-peptide complex with low density lipoprotein receptor-related protein-2/megalin. *J. Biol. Chem.* **272**, 18644–18649.

Han, B. H., DeMattos, R. B., Dugan, L. L., Kim-Han, J. S., Brendza, R. P., Fryer, J. D., Kierson, M., Cirrito, J. R., Quick, K., Harmony, J. A. K., Aronow, B. J., and Holtzman, D. M. (2001). Clusterin contributes to caspase-3-independent brain injury following neonatal hypoxia-ischemia. *Nat. Med.* **7**, 338–343.

Hatters, D. M., Wilson, M. R., Easterbrook-Smith, S. B., and Howlett, G. J. (2002). Suppression of apolipoprotein C-II amyloid formation by the extracellular chaperone, clusterin. *Eur. J. Biochem.* **269**, 2789–2794.

Humphreys, D., Hochgrebe, T. T., Easterbrook-Smith, S. B., Tenniswood, M. P. R., and Wilson, M. R. (1997). Effects of clusterin overexpression on TNFalpha- and TGFbeta-mediated death of L929 cells. *Biochemistry* **36**, 15233–15243.

Humphreys, D. T., Carver, J. A., Easterbrook-Smith, S. B., and Wilson, M. R. (1999). Clusterin has chaperone-like activity similar to that of small heat shock proteins. *J. Biol. Chem.* **274**, 6875–6881.

Jenne, D. E., and Tschopp, J. (1989). Molecular structure and functional characterization of a human complement cytolysis inhibitor found in blood and seminal plasma: Identity to sulphated glycoprotein 2, a constituent of rat testis fluid. *Proc. Natl. Acad. Sci. USA* **86**, 7123–7127.

Jensen, P. E., and Sottrup-Jensen, L. (1986). Primary structure of human alpha-2 macroglobulin. Complete disulfide bridge assignment and localization of two interchain bridges in the dimeric and proteinase binding unit. *J. Biol. Chem.* **261**, 15863–15869.

Kang, S. W., Shin, Y. J., Shim, Y. J., Jeong, S. Y., Park, I. S., and Min, B. H. (2005). Clusterin interacts with SCLIP (SCG10-like protein) and promotes neurite outgrowth of PC12. *Exp. Cell Res.* **309**, 305–315.

Kida, E., Choi-Miura, N. H., and Wisniewski, K. E. (1995). Deposition of apolipoproteins E and J in senile plaques is topographically determined in both Alzheimer' disease and Down' syndrome brain. *Brain Res.* **685**, 211–216.

Kirszbaum, L., Sharpe, J. A., Murphy, B., d'Apice, A. J., Classon, B., Hudson, P., and Walker, I. D. (1989). Molecular cloning and characterization of the novel, human complement-associated protein, SP-40,40: A link between the complement and reproductive systems. *EMBO J.* **8**, 711–718.

Kliffen, M., de Jong, P. T., and Luider, T. M. (1995). Protein analysis of human maculae in relation to age-related maculopathy. *Lab. Invest.* **72**, 267–272.

Kounnas, M. Z., Loukinova, E. B., Steffansson, S., Harmony, J. A. K., Brewer, B. H., Strickland, D. K., and Argraves, W. S. (1995). Identification of glycoprotein 330 as an endocytic receptor for appolipoprotein J/Clusterin. *Biochemistry* **270**, 13070–13075.

Kumita, J. R., Poon, S., Caddy, G. L., Hagan, C. L., Dumoulin, M., Yerbury, J. J., Stewart, E. M., Robinson, C. V., Wilson, M. R., and Dobson, C. M. (2007). The extracellular chaperone clusterin potentially inhibits amyloid formation by interacting with prefibrillar species. *J. Mol. Biol.* **369**, 157–167.

Kurosky, A., Barnett, D. R., Lee, T. H., Touchstone, R. E., Hay, R. E., Arnott, M. S., Bowman, B. H., and Fitch, W. M. (1980). Covalent structure of human haptoglobin: A serine protease homolog. *Proc. Natl. Acad. Sci. USA* **77**, 3388–3392.

Lakins, J. N., Poon, S., Easterbrook-Smith, S. B., Carver, J. A., Tenniswood, M. P. R., and Wilson, M. R. (2002). Evidence that clusterin has discrete chaperone and ligand binding sites. *Biochemistry* **41**, 282–291.

Langlois, M. R., and Delanghe, J. R. (1996). Biological and clinical significance of haptoglobin polymorphisms in humans. *Clin. Chem.* **42**, 1589–1600.

Lauer, D., Reichenbach, A., and Birkenmeier, G. (2001). Alpha 2-macroglobulin-mediated degradation of amyloid beta 1–42: A mechanism to enhance amyloid beta catabolism. *Exp. Neurol.* **169**, 385–392.

Law, G. L., and Griswold, M. D. (1994). Activity and formof sulphated glycoprotein 2 (clusterin) from cultured sertoli cells, testis, and epididymus of the rat. *Biol. Reprod.* **50**, 669–679.

Lee, G. J., Roseman, A. M., Saibil, H. R., and Vierling, E. (1997). A small heat shock protein stably binds heat-denatured model substrates and can maintain a substrate in a folding-competent state. *EMBO J.* **16**, 659–671.

Leskov, K. S., Klokov, D. Y., Li, J., Kinsella, T. J., and Boothman, D. A. (2003). Synthesis and functional analyses of nuclear clusterin, a cell death protein. *J. Biol. Chem.* **278**, 11590–11600.

Lim, S. K., Kim, H., Bin Ali, A., Lim, Y. K., Wang, Y., Chong, S. M., Costantini, F., and Baumman, H. (1998). Increased susceptibility in Hp knockout mice during acute hemolysis. *Blood* **92**, 1870–1877.

Loison, F., Debure, L., Nizard, P., Le Goff, P., Michel, D., and Le Drean, Y. (2006). Up-regulation of the clusterin gene after proteotoxic stress: Implications of HSF1–HSF2 hetero-complexes. *Biochem. J.* **395**, 223–231.

Lopez de la Paz, M., and Serrano, L. (2004). Sequence determinants of amyloid fibril formation. *Proc. Natl. Acad. Sci. USA* **101**, 87–92.

Lundgren, S., Carling, T., Hjälm, G., Juhlin, C., Rastad, J., Pihlgren, U., Rask, L., Åkerström, G., and Hellman, G. (1997). Tissue distribution of human gp330/megalin, a putative Ca 2 + sensing protein. *J. Histochem. Cytochem.* **45**, 383–392.

Mahon, M. G., Linstedt, K. A., Hermann, M., Nimpf, J., and Schneider, W. J. (1999). Multiple involvement of clusterin in chicken ovarian follicle development. *J. Biol. Chem.* **274**, 4036–4044.

Makin, O. S., Atkins, E., Sikorski, P., Johansson, J., and Serpell, L. C. (2005). Molecular basis for amyloid fibril formation and stability. *Proc. Natl. Acad. Sci. USA* **102**, 315–320.

Margineanu, I., and Ghetie, V. (1981). A selective model of plasma protein catabolism. *J. Theor. Biol.* **90**, 101–110.

Marini, I., Moschini, R., Corso, A. D., and Mura, U. (2005). Chaperone-like features of bovine serum albumin: A comparison with alpha-crystalin. *Cell. Mol. Life Sci.* **62**, 3092–3099.

Matsubara, E., Frangione, B., and Ghiso, J. (1995). Characterization of apolipoprotein J-Alzheimer's a-beta interaction. *J. Biol. Chem.* **270**, 7563–7567.

Matt-Schieman, M. L., van Duinen, S. G., Bornebroek, M., Haan, J., and Roos, R. A. (1996). Hereditary cerebral hemorrhage with amyloidosis-Dutch type (HCHWA-D): II-A review of histopathological aspects. *Brain Pathol.* **6,** 115–120.

McGeer, E. G., and McGeer, P. L. (1997). Innate inflammatory reaction of the brain in Alzheimer disease. *McGill J. Med.* **3,** 134–144.

McHattie, S., and Edington, N. (1999). Clusterin prevents aggregation of neuropeptide 106–126 *in vitro. Biochem. Biophys. Res. Commun.* **259,** 336–340.

McLaughlin, L., Zhu, G., Mistry, M., Ley-Ebert, C., Stuart, W. D., Florio, C. J., Groen, P. A., Witt, S. A., Kimball, T. R., Witte, D. P., Harmony, J. A. K., and Aronow, B. J. (2000). Apolipoprotein J/clusterin limits the severity of murine autoimmune myocarditis. *J. Clin. Invest.* **106,** 1105–1113.

Michel, D., Chatelain, G., North, S., and Brun, G. (1997). Stress-induced transcription of the clusterin/apoJ gene. *Biochem. J.* **328,** 45–50.

Motomiya, Y., Ando, Y., Haraoka, K., Sun, X., Iwamoto, H., Uchimura, T., and Maruyama, T. (2003). Circulating levels of alpha2-macroglobulin-beta2-microglobulin complex in hemodialysis patients. *Kidney Int.* **64,** 2244–2252.

Murphy, B. F., Kriszbaum, L., Walker, I. D., and d'Apice, J. F. (1988). SP-40,40, a newly identified normal human serum protein found in the SC5b-9 complex of complement and in the immune deposits in glomerulonephritis. *J. Clin. Invest.* **81,** 1858–1864.

Nishida, K., Quantock, A. J., Dota, A., Choi-Mura, N. H., and Kinoshita, S. (1999). Apolipoprotein J and E co-localise with amyloid in gelatinous drop-like and lattice type I corneal dystrophies. *Br. J. Opthalmol.* **83,** 1178–1182.

Nizard, P., Tetley, S., Le Drean, Y., Watrin, T., Le Goff, P., Wilson, M. R., and Michel, D. (2007). Stress-induced retrotranslocation of clusterin/ApoJ into the cytosol. *Traffic* **8,** 554–565.

Oda, T., Osterberg, H. H., Johnson, S. A., Pasinetti, G. M., Morgan, T. E., Rozovsky, I., Stine, W. B., Snyder, S. W., and Holzman, T. F. (1995). Clusterin (apoJ) alters the aggregation of amyloid beta peptide 1–42 and forms slowly sedimenting A-beta complexes that cause oxidative stress. *Exp. Neurol.* **136,** 22–31.

Park, D. C., Yeo, S. G., Wilson, M. R., Yerbury, J. J., Kwon, J., and Welch, W. R. (2008). Clusterin interacts with Paclitaxel and confer Paclitaxel resistance in ovarian cancer. *Neoplasia* **10,** 964–972.

Pavlicek, Z., and Ettrich, R. (1999). Chaperone-like activity of human haptoglobin: Similarity with a-crystallin. *Collect. Czech. Chem. Commun.* **64,** 717–725.

Poon, S., Rybchyn, M. S., Easterbrook-Smith, S. B., Carver, J. A., and Wilson, M. R. (2000). Clusterin is an ATP-independent chaperone with a very broad substrate specificity that stabilizes stressed proteins in a folding-competent state. *Biochemistry* **39,** 15953–15960.

Poon, S., Rybchyn, M. S., Easterbrook-Smith, S. B., Carver, J. A., Pankhurst, G. J., and Wilson, M. R. (2002a). Mildly acidic pH activates the extracellular molecular chaperone clusterin. *J. Biol. Chem.* **277,** 39532–39540.

Poon, S., Treweek, T. M., Wilson, M. R., and Easterbrook-Smith, S. B. (2002b). Clusterin is an extracellular chaperone that specifically interacts with slowly aggregating proteins on their off-folding pathway. *FEBS Lett.* **513,** 259–266.

Powers, J. M., Schlaepfer, W. W., Willingham, M. C., and Hall, B. J. (1981). An immunoperoxidase study of senile cerebral amyloidosis with pathogenetic considerations. *J. Neuropathol. Exp. Neurol.* **40,** 592–612.

Reddy, K. B., Jin, G., Karode, M. C., Harmony, J. A. K., and Howe, P. H. (1996). Transforming growthfactor b (TGFb)-induced nuclear localization of apolipoprotein J/clusterin in epithelial cells. *Biochemistry* **35,** 6157–6163.

Rosenberg, M. E., and Silkensen, J. (1995). Clusterin: Physiologic and pathophysiologic considerations. *Int. J. Biochem. Cell Biol.* **27,** 633–645.

Rosenberg, M. E., Girton, R., Finkel, D., Chmielewski, D., Barrie, A., Witte, D. P., Zhu, G., Bisslcr, J. J., Harmony, J. A. K., and Aronow, B. J. (2002). Apolipoprotein J/clusterin prevents progressive glomerulopathy of aging. *Mol. Cell. Biol.* **22**, 1893–1902.

Sakaguchi, H., Takeya, M., Suzuki, H., Hakamata, H., Kodama, T., Horiuchi, S., Gordon, S., van der Laan, L. J., Ishibashi, S., Kitamura, N., and Takahashi, K. (1998). Role of macrophage scavenger receptors in diet-induced atherosclerosis in mice. *Lab. Invest.* **78**, 423–434.

Santilli, G., Aronow, B. J., and Sala, A. (2003). Essential requirement of apolipoprotein J (clusterin) signalling for Ikappa B expression and regulation of NF-kappaB activity. *J. Biol. Chem.* **278**, 38214–38219.

Sasaki, K., Doh-ura, K., Wakisaka, Y., and Iwaki, T. (2002). Clusterin/apolipoprotein J is associated with cortical Lewy bodies: Immunohistochemical study in cases with alpha-synucleinopathies. *Acta Neuropathol.* **104**, 225–230.

Shibata, M., Yamada, S., Ram Kumar, S., Calero, M., Bading, J., Frangione, B., Holtzman, D. M., Miller, C. A., Strickland, D. K., Ghiso, J., and Zlokovic, B. V. (2000). Clearance of Alzheimer's amyloid-ss(1–40) peptide from brain by LDL receptor-related protein-1 at the blood–brain barrier. *J. Clin. Invest.* **106**, 1489–1499.

Sobek, O., and Adam, P. (2003). On S. Seyfert, V. Kunzmann, N. Schwetfeger, H.C. Koch, A. Faulstich: Determinants of lumbar CSF protein concentration. *J. Neurol.* **250**, 371–372.

Sottrup-Jensen, L. (1989). Alpha-macroglobulins: Structure shape and mechanism of proteinase complex formation. *J. Biol. Chem.* **264**, 11539–11542.

Strivastava, P. (2002). Roles of heat-shock proteins in innate and adaptive immunity. *Nat. Rev. Immunol.* **2**, 185–194.

Strocchi, P., Smith, M. A., Perry, G., Tamagno, E., Danni, O., Pession, A., Gaiba, A., and Dozza, B. (2006). Clusterin up-regulation following sub-lethal oxidative stress and lipid peroxidation in human neuroblastoma cells. *Neurobiol. Aging* **27**, 1588–1594.

Tang, H., Fu, Y., Cui, Y., He, Y., Zeng, X., Ploplis, V. A., Castellino, F. J., and Luo, Y. (2009). Fibrinogen has chaperone-like activity. *Biochem. Biophys. Res. Commun.* **378**, 662–667.

Trougakos, I. P., Pawelec, G., Tzavelas, C., Ntouroupi, T., and Gonos, E. S. (2006). Clusterin/apolipoprotein J up-regulation after zinc exposure, replicative senescence or differentiation of human haematopoietic cells. *Biogerontology* **7**, 375–382.

Trougakos, I. P., Lourda, M., Antonelou, M. H., Kletsas, D., Gorgoulis, V. G., Papassideri, I. S., Zou, Y., Margaritis, L. H., Boothman, D., and Gonos, E. S. (2009). Intracellular clusterin inhibits mitochondrial apoptosis by suppressing p53-activating stress signals and stabilizing the cytosolic Ku70–Bax protein complex. *Clin. Cancer Res.* **15**, 48–59.

Ubrich, C., Fritzenwanger, M., Zeiher, A. M., and Dimmeler, S. (2000). Laminar shear stress upregulates the complement-inhibitory protein clusterin. *Circulation* **101**, 352–355.

Wehrli, P., Charnay, Y., Vallet, P., Zhu, G., Harmony, J., Aronow, B., Tschopp, J., Bouras, C., Viard-Leveugie, I., French, L. E., and Giannakopoulos, P. (2001). Inhibition of post-ischemic brain injury by clusterin overexpression. *Nat. Med.* **7**, 977–978.

Wicher, G. K., and Aldskogius, H. (2005). Adult motor neurons show increased susceptibility to axotomy-induced death in mice lacking clusterin. *Eur. J. Neurosci.* **21**, 2024–2028.

Wickner, S., Maurizi, M. R., and Gottesman, S. (1999). Posttranslational quality control: Folding, refolding, and degrading proteins. *Science* **286**, 1888–1893.

Wilson, M. R., and Easterbrook-Smith, S. B. (1992). Clusterin binds by a multivalent mechanism to the Fc and Fab regions of IgG. *Biochim. Biophys. Acta* **1159**, 319–326.

Wilson, M. R., and Easterbrook-Smith, S. B. (2000). Clusterin is a secreted mammalian chaperone. *Trends Biochem. Sci.* **25**, 95–98.

Witte, D. P., Aronow, B. J., Stauderman, M. L., Stewart, W. D., Clay, M. A., Gruppo, R. A., Jenkins, S. H., and Harmony, J. A. (1993). Platelet activation releases megakaryocyte-synthesized apolipoprotein J, a highly abundant protein in a atheromatous lesions. *Am. J. Pathol.* **143**, 763–773.

Wood, S. J., Chan, W., and Wetzel, R. (1996). Seeding of Abeta fibril formation is inhibited by all three isotypes of apolipoprotein E. *Biochemistry* **35**, 12623–12628.

Yang, C. R., Leskov, K., Hosley-Eberlein, K., Criswell, T., Pink, J. J., Kinsella, T. J., and Boothman, D. A. (2000). Nuclear clusterin/XIP8, an x-ray induced Ku70-binding protein that signals cell death. *Proc. Natl. Acad. Sci. USA* **97**, 5907–5912.

Yerbury, J. J., Rybchyn, M. S., Easterbrook-Smith, S. B., Henriques, C., and Wilson, M. R. (2005a). The acute phase protein haptoglobin is a mammalian extracellular chaperone with an action similar to clusterin. *Biochemistry* **44**, 10914–10925.

Yerbury, J. J., Stewart, E., Wyatt, A. R., and Wilson, M. R. (2005b). Quality control of protein folding in extracellular space. *EMBO Reports* **6**, 1131–1136.

Yerbury, J. J., Poon, S., Meehan, S., Thompson, B., Kumita, J. R., Dobson, C. M., and Wilson, M. R. (2007). The extracellular chaperone clusterin influences amyloid formation and toxicity by interacting with pre-fibrillar structures. *FEBS Lett.* **21**, 2312–2322.

Yerbury, J. J., Kumita, J. R., Meehan, S., Dobson, C. M., and Wilson, M. R. (2009). Alpha 2 macroglobulin and haptoglobin supress amyloid formation by interacting with prefibrillar protien species. *J. Biol. Chem.* **284**, 4246–4254.

Zenkel, M., Kruse, F. E., Junemann, A. G., Naumann, G. O., and Schlotzer-Schrehardt, U. (2006). Clusterin deficiency in eyes with pseudoexfoliation syndrome may be implicated in the aggregation and deposition of pseudoexfoliative material. *Invest. Opthalmol. Vis. Sci.* **47**, 1982–1990.

Zhang, H. L., Kim, J. K., Edwards, C. A., Xu, Z. H., Taichman, R., and Wang, C. Y. (2005). Clusterin inhibits apoptosis by interacting with activated Bax. *Nat. Cell Biol.* **7**, 909–915.

Zlokovic, B. V., Martel, C. L., Matsubara, E., McComb, J. G., Zheng, G., McCluskey, R. T., Frangione, B., and Ghiso, J. (1996). Glycoprotein 330 megalin: Probable role in receptor-mediated transport of apolipoprotein J alone and in a complex with Alzheimer disease amyloid b at the blood–brain and blood–cerebrospinal fluid barriers. *Proc. Natl. Acad. Sci. USA* **93**, 4229–4234.

Cell Protective Functions of Secretory Clusterin (sCLU)

Gerd Klock, Markus Baiersdörfer, and Claudia Koch-Brandt

Institute of Biochemistry, Joh.-Gutenberg University of Mainz, Becherweg 30, D-55099 Mainz, Germany

Secretory clusterin (sCLU) is found as an 80-kDa glycoprotein in virtually all body fluids, in serum it is associated with high-density lipoprotein (HDL). Here, we discuss demonstrated and proposed mechanisms of the cytoprotective functions of sCLU in instances of apoptosis, necrosis, and disease. These include prevention from cell damage by lipid oxidation in blood vessels, removal of dead cell remnants in tissues undergoing various forms of cell death, and clearance of harmful extracellular molecules such as amyloid beta (Aβ) by endocytosis or transcytosis. All these functions may reflect the propensity of sCLU to bind to a wide spectrum of hydrophobic molecules on one hand and to specific cell-surface receptors on the other hand. Identified and proposed sCLU receptors are members of the LDL receptor family of endocytosis receptors. Since these receptors recently have proved to modulate cell signaling we will discuss whether sCLU due to this interaction not only targets its ligands for clearance, but may also be involved in triggering signal transduction. © 2009 Elsevier Inc.

I. INTRODUCTION

Clusterin (CLU) has surfaced repeatedly by virtue of its overexpression in the face of tissue remodeling and degeneration (Koch-Brandt and Morgans, 1996). It is upregulated in model systems of apoptosis and injury, in malignant tumors, in epileptic foci, and in the hippocampus of M. Alzheimer patients. The association of CLU gene expression with stress conditions and apoptosis, the tightly regulated "programmed" cell death program, has suggested that the protein could be involved in the cell death process. Depending on cell type and experimental conditions, CLU isoforms were

Advances in CANCER RESEARCH
Copyright 2009, Elsevier Inc. All rights reserved.

reported to favor either cell death or cell survival, activities with an impact on many diseases such as cancer, autoimmune, and cardiovascular diseases. In this chapter, we will concentrate on secretory clusterin (sCLU), the most abundant isoform, with a focus on its protective functions in apoptotic and necrotic tissues.

II. sCLU—A COMPONENT OF HIGH-DENSITY LIPOPROTEINS

There is a broad agreement today concerning the risk of a high concentration of low-density lipoprotein (LDL) and LDL cholesterol to developing atherosclerotic diseases, while high-density lipoprotein (HDL) and HDL cholesterol are considered to be protective (Gordon and Rifkind, 1989). A number of models exist to explain the protection by an increased HDL concentration, including improved reverse cholesterol transport and cholesterol efflux from macrophages, the cell type that is believed to develop a critical phenotype following ingestion of modified LDL particles named "foam cells" (for review, see deGoma *et al.*, 2008). Furthermore, HDL was reported to have anti-inflammatory, antithrombotic, and antioxidative functions (Navab *et al.*, 1991; van Lenten *et al.*, 1995; Watson *et al.*, 1995). sCLU (also designated apolipoprotein J in this context) was detected associated with HDL subfractions (de Silva *et al.*, 1990; Jenne *et al.*, 1991) which suggested that it could be protective in the presence of proatherogenic conditions.

One possible mechanism of the protective effect of sCLU, either alone or as a component of HDL was concluded by cell-culture experiments where cells were incubated with mildly oxidized LDL in the presence or absence of sCLU. The "minimally oxidized LDL" stimulated both lipid peroxide formation and monocyte migration, and the presence of sCLU was able to prevent both activities of the modified LDL (Navab *et al.*, 1997). On the basis of these results, sCLU was suggested to have protective activity and prevent the pathological consequences of LDL modification in the blood vessels.

In *in vivo* studies in animals, the relative expression of the two plasma proteins sCLU and paraoxonase (PON), another HDL-associated protein, was found to change in response to atherosclerotic conditions. Their relative level (sCLU/PON) increased, that is, sCLU increased while PON decreased, in the following animal models (i) in apoE knockout mice when compared to wild-type mice (9-fold increase), and (ii) in LDL receptor (LDLR) knockout mice that were on a cholesterol-enriched diet (over 100-fold increase) (Navab *et al.*, 1997). Since both proteins display protective activities, the upregulation of CLU could constitute a reaction to the pathological conditions to minimize the disease, while the downregulation of PON could be causative for the progression of the disease. The observation, that a higher sCLU/PON ratio correlates with lowered HDL protective activity (Navab et al., 1997), is in line with the present consensus that PON is the major contributor to the beneficial effect of HDL.

In contrast to these animal studies, an analysis of human sCLU serum levels showed a positive correlation of sCLU with PON both in men and women (Kujiraoka et al., 2006). Furthermore, in the serum of male coronary heart disease patients, after adjustment for covariants (PON; triglycerides and apo-A-II, both negatively correlated), sCLU was significantly lower than in healthy controls, which may point to an antiatherogenic sCLU activity also in the human circulation (Kujiraoka et al., 2006).

Recently, the beneficial activity of sCLU protein has been localized to small peptides derived from the sequence of mature secreted clusterin which had been predicted to form amphipathic helices. The peptides tested in in vivo experiments, which were composed of either L- or D-amino acids improved both cholesterol export from cells into plasma and HDL anti-inflammatory properties (Navab et al., 2005, 2007). The protective effect of sCLU and peptides derived from it could be due to its sequestering activity of harmful substances in order to avoid attracting immune cells and inducing atherosclerotic changes in the vessel wall. In line with the model that sequestering LDL-related modified lipoproteins could contribute to the protective effect of sCLU, protection from the cytotoxic effects of enzymatically modified low-density lipoprotein (E-LDL) paralleled direct sCLU binding to the lipoprotein (Schwarz et al., 2008).

III. sCLU IN APOPTOSIS—SIGNALING TOWARD CELL SURVIVAL?

Apoptosis (or programmed cell death) is a normal process that allows tissue remodeling during development and, by counteracting proliferation, limits uncontrolled cell growth and preserves tissue homeostasis. This highly ordered cell death pathway ensures that cellular components of dying cells are deposited in membrane-covered compartments (apoptotic vesicles) before they are ingested by phagocytic cells. This process and its tight regulation avoid the uncontrolled release of cellular components and helps prevent inadequate immune responses. In contrast, necrotic cell death results from external noxes, such as mechanical or chemical damage, or pathogen infection. Necrosis is a process that by definition results in material leaking from the dying cells that can stimulate the innate immune system and result in inflammation. In addition, necrotic material could eventually elicit a specific adaptive immune response to cellular antigens resulting in an auto-immune reaction, whereas apoptotic cells may produce immunosuppressive signals avoiding an immune response (Frey *et al.*, 2008; Patel *et al.*, 2006). In tumors, apoptosis and necrosis have frequently been reported and may be induced in poorly vasculated areas of solid tumors (Greijer and van der Wall, 2004; Zhou *et al.*, 2006).

CLU gene expression is upregulated in tumors of various origins (Ahn *et al.*, 2008; Chou *et al.*, 2009; Kadomatsu *et al.*, 1993; Parczyk *et al.*, 1994; Redondo *et al.*, 2000; Wellmann *et al.*, 2000; for review, see Shannan *et al.*, 2006). It is further increased under conditions known to induce apoptosis, such as hormone withdrawal in estrogen-dependent breast tumor, or in an-drogen-dependent prostate cancer models (Brändström *et al.*, 1994; Chen *et al.*, 1996; Kyprianou *et al.*, 1991) and other conditions involving cell death, such as tissue remodeling and injury (Bandyk *et al.*, 1990; Bursch *et al.*, 1995; Buttyan *et al.*, 1989). There are two straightforward models to explain how the CLU gene might be regulated in tumors: (a) by oncogene activation causing an altered intracellular signaling resulting in the activation of the CLU gene, or (b) by the apoptotic and/or necrotic microenvironment in tumor tissue with the exposition or release of signaling molecules that modify CLU gene expression. We have demonstrated in Rat-1 fibroblast cells stably transfected with either ras or myc or both oncogenes, that neither oncogene alone nor a ras–myc combination caused an induction of the clusterin gene (Klock *et al.*, 1998) arguing against a direct role of the oncogenes in the control of CLU gene expression. In support of a role of the microenvironment, components exposed or released by apoptotic/necrotic cells were demon-strated to produce clusterin gene upregulation. Similarly, apoptosis provoked by androgen withdrawal to treat prostate carcinoma (Brändström *et al.*, 1994), or cell death in tumors or normal tissue induced by conditions such as hypoxia could result in the release of signals that activate the CLU gene in the vital bystander cells (Kim *et al.*, 2007; Poulios *et al.*, 2006). In fact, it has been demonstrated that, upon induction of apoptosis in different tumor cell types, including carcinoma and lymphoma cells, and in aging neutrophils CLU mRNA expression was low or undetectable in the dying cells, but was specifically induced in the surviving cells (French *et al.*, 1994). These results supported the notion that the CLU gene product, presumably sCLU, was not actively promoting apoptosis but rather could play a role in protecting the vital bystander cells. This contrasts with reports on intracellular CLU iso-forms proposed to causally exert proapoptotic or antiapoptotic functions in tumor cells (Ammar and Closset, 2008; Bailey *et al.*, 2002; Caccamo *et al.*, 2003; Chen *et al.*, 2004; Leskov *et al.*, 2003; Miyake *et al.*, 2000; Moretti *et al.*, 2007; Ranney *et al.*, 2007; Trougakos and Gonos, 2002; Trougakos *et al.*, 2009; Zhang *et al.*, 2005; see accompanying chapters in this volume).

In contrast to the complex situation with regard to the intracellular isoforms, sCLU has now been unambiguously found to be cytoprotective in the presence of apoptosis and necrosis (Miyake *et al.*, 2000; Sensibar *et al.*, 1995; Shannan *et al.*, 2006; Sintich *et al.*, 1999; You *et al.*, 2003). One possible mechanism how sCLU exerts its protective function in the affected tissues involves binding of the protein to a cellular receptor. Megalin, the first identified sCLU receptor (Bartl *et al.*, 2001; Farquhar *et al.*, 1995; Kounnas, *et al.*, 1995) belongs to the LDL-receptor (LDLR) family of structurally related membrane receptors (for review, see May *et al.*, 2005). These receptors bind ligands with a broad specificity, and internalize them by endocytosis. Ligands binding to these receptors include lipoproteins such as LDL in the case of the LDLR as well as various proteins like protease inhibitors such as serpin A1/alpha1-antitrypsin and the corresponding inhibitor/protease complexes in the case of LRP-1 (Strickland and Ranganathan, 2003; Strickland *et al.*, 1995). Interestingly, ligand binding to these receptors may elicit signal transduction, as has been shown for the VLDL receptor (VLDLR), the apoE receptor 2 (apoER2), LRP-1, and mega-lin (LRP-2) (for review, see May *et al.*, 2005). Remarkably, evidence has been presented concluding that sCLU interacts with all these receptors (Bajari *et al.*, 2003; Bartl *et al.*, 2001; Kounnas, *et al.*, 1995).

The mechanistic details how sCLU could modulate cell signaling upon receptor binding was analyzed by CLU overexpression using a doxycyclin-inducible (Tet-on) promoter in the prostate cell line MLL. The cells were protected from TNF-α-induced apoptosis by induction of sCLU expression (Ammar and Closset, 2008). Conditioned medium from these cells contain-ing sCLU, or the purified protein, both resulted in cytoprotection. Interest-ingly, in this study, evidence was presented pointing to sCLU/megalin interaction-induced signaling as the cause of cell protection: firstly, follow-ing CLU gene induction, a sCLU–megalin complex is formed as revealed by immune precipitation and Western blot experiments; secondly, clusterin expression led to increased phosphorylation of megalin and Akt; and thirdly, increased phosphorylation of both, megalin and Akt, could be inhibited by the PI3 kinase inhibitor wortmannin, indicating that sCLU binding to its receptor megalin activated the PI3 kinase/Akt pathway and produced multi-ple protein phosphorylation. However, in the same study, inhibition of this pathway did not block the protective effect which indicated that additional pathways may be involved in the protection by sCLU-induced signaling (Ammar and Closset, 2008).

In this context, by binding to growth factor IGF1, sCLU was reported to block its interaction with the IGF receptor, thereby preventing receptor signaling including PI3 kinase/Akt pathway activation (Jo *et al.*, 2008). Intriguingly, the interaction of sCLU with IGF1 produces an effect on the PI3 kinase/Akt pathway contrary to that resulting from the sCLU/megalin interaction. Therefore these studies imply that, depending on its interacting partner, sCLU serves as a trigger that controls the activity of the PI3 kinase/Akt pathway in either way.

Most recently, it was shown that sCLU stimulates astrocyte proliferation by modulating EGF receptor signaling activity, resulting in MEK/ERK stimulation; and the possibility that this stimulation may occur via cross talk with a sCLU receptor, such as megalin, was discussed (Shim *et al.*, 2009). The same pathway was affected in a human adenocarcinoma cell line where CLU expression had been knocked down, however, in this study the active CLU isoform had not been characterized (Chou *et al.*, 2009).

In addition to megalin, other lipoprotein receptors, namely LRP-1, which has been shown to regulate PDGF signaling in the vascular wall (Boucher *et al.*, 2002), to limit expression of inflammatory genes *in vivo* and *in vitro* (Zurhove *et al.*, 2008), to mediate Wnt signaling (Terrand *et al.*, 2009), and to interrupt, by ligand binding, IL-1β signaling in vascular smooth muscle cells (VSMCs) (Kawamura *et al.*, 2007), may bind sCLU and activate signal transduction pathways (Bartl *et al.*, 2001). Furthermore, in brain development, two sCLU receptor candidates, ApoER2 and VLDLR, bind to their ligand reelin which leads to phosphorylation of Disabled-1 (D'Arcangelo *et al.*, 1999; Hiesberger *et al.*, 1999; Trommsdorff *et al.*, 1999). Finally, both LRP-1 and apoER2 bind to FE65, a signaling adaptor protein active in neuronal development (McLoughlin and Miller, 2008). These data open intriguing perspectives with regard to potential signaling functions of sCLU. However, whether sCLU elicits signal transduction by binding to any of the candidates of the LDLR family, has yet to be shown.

Along these lines, a contribution to development and/or progression of certain cancers was discussed for individual receptors. In the case of the megalin gene, a correlation of single-nucleotide polymorphisms with prostate cancer recurrence/progression and with mortality was reported (Holt *et al.*, 2008), and the cancer progression-associated ligand midkine was reported to bind to LRP-1 (Chen *et al.*, 2007). Some of these observations could be interpreted as consistent with sCLU binding to these receptors, thereby promoting tumor cell survival. Again, however, whether specific sCLU/receptor interactions indeed modify cancer development, progression, and invasion remains to be analyzed.

IV. sCLU IN THE REMOVAL OF DEAD CELLS AND CELLULAR DEBRIS

The association of sCLU expression with cellular stress and cell death processes suggested that the protein may play a role in homeostasis. Two functions were assigned to sCLU which may contribute to this task;

firstly, the chaperone-like activity which keeps denatured proteins in solution (Wilson and Easterbrook-Smith, 2000); and secondly, its function as a ligand of LDLR family proteins which are involved in the cellular uptake processes endocytosis and phagocytosis, and in some instances in signal transduction.

Generally, cells that die by apoptosis or by necrosis, and their remnants are removed by phagocytosis. In addition to "professional" phagocytes, macrophages, so-called "nonprofessional" phagocytes such as epithelial cells, fibroblasts, or smooth muscle cells will aid this process, especially when a large number of dead cells accumulate within a short time (Erwig and Henson, 2008; Fries *et al.*, 2005; Henson and Hume, 2006; Platt *et al.*, 1998).

The association of the expression of sCLU with cell death raised the question whether the protein could itself be actively involved in the removal of denatured proteins, apoptotic cells, and debris. In general, "helper" proteins may be required by phagocytic cells, specifically by nonprofessional phagocytes, for the ingestion of dying cells and apoptotic vesicles, by acting locally as opsonizing proteins to support binding and uptake of cell remnants and debris (Hart *et al.*, 2004; Lillis *et al.*, 2008; Vandivier *et al.*, 2002a).

In fact, an active role for sCLU in phagocytosis was suggested when fibroblasts and epithelial cells were tested as nonprofessional phagocytes (Bartl *et al.*, 2001). In LLC-PK1 cells, an epithelial cell of renal proximal tubule, sCLU was found to bind to debris, both from apoptotic and necrotic cells and stimulate their uptake, while the protein was itself internalized and partially degraded (Fig. 1). In contrast, the macrophage cell line J774 did not support sCLU-dependent uptake of cell debris (Bartl *et al.*, 2001), which could be due to the presence of more efficient disposal systems involving scavenger receptors in this cell type (Józefowski *et al.*, 2005). Phagocytosis was reduced when either megalin, or the related receptor LRP-1 were pharmacologically blocked or genetically inactivated, demonstrating that sCLU in cooperation with these receptors stimulates phagocytic uptake of cell debris (Bartl *et al.*, 2001). sCLU affinity to both megalin and LRP-1 may ensure uptake of cell debris on both sides of epithelial cell layers, since megalin and LRP-1 are located on the apical and basolateral plasma membrane domain, respectively (Bartl *et al.*, 2001; Marzolo *et al.*, 2003).

Apoptotic and necrotic cells display signals on their surface which are recognized by phagocytic cells (Erwig and Henson, 2008; Platt *et al.*, 1998; Züllig and Hengartner, 2004). A well-established signal is the lipid phosphatidylserine (PS), which during apoptosis appears on the outer leaflet of the lipid bilayer of the plasma membrane, thereby serving as a marker of dying cells versus vital cells. A number of receptor candidates were reported to be involved in PS recognition during the process of phagocytosis (for review, see Savill and Gregory, 2007): PSR (Fadok *et al.*, 2000), whose significance is still under discussion (Kolb *et al.*, 2007; Mitchell *et al.*, 2006); TIM-4 (Miyanishi *et al.*, 2007); stabilin-2 (Jeannin *et al.*, 2008); and BAI1

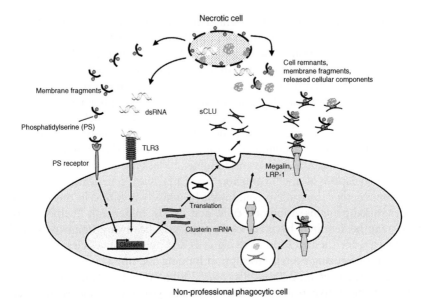

Fig. 1 A model depicting the proposed regulation and function of sCLU in the face of apoptosis and necrosis. Components of apoptotic and necrotic cells, exposed phosphatidylserine (PS), and double-stranded RNA (dsRNA) released from necrotic cells, induce clusterin gene expression. The secreted clusterin binds cell debris and targets it to a receptor of the LDL receptor family, such as megalin or LRP-1 leading to the internalization of the complex ligand followed by its degradation and recycling of the receptor. (See Page 7 in Color Section at the back of the book.)

(Park *et al.*, 2007). Remarkably, both apoptotic and necrotic cell debris were found to induce CLU gene expression (Bach *et al.*, 2001; Baiersdörfer *et al.*, manuscript submitted for publication). Furthermore, by incubating healthy cells with liposomes of defined lipid composition, it was demonstrated that CLU mRNA induction relies on PS, which is exposed on both apoptotic bodies and necrotic cell debris (Bach *et al.*, 2001), showing that this lipid acts as a signaling molecule to modify gene expression in the surviving nonphagocytic cells to induce a component of the disposal system which is not recycled but degraded upon action (see Fig. 1).

The notion, that sCLU fulfills an important function in homeostasis by promoting the uptake of dead cells and cell remnants, gained support, when clusterin knockout- and wild-type mice were compared. Following heat treatment, the disappearance of affected germ cells, which was observed in the wild-type mice, showed a delay for about 1 day in the knockout mice, suggesting that the inactivation of the CLU gene results in an impaired uptake of dead cells and/or cell debris in the reproductive system (Bailey *et al.*, 2002). These observations underline the function of sCLU in the removal process of dead cells, specifically by nonprofessional phagocytes;

and it could provide a rationale for the observed upregulation of the gene in the presence of cell remnants (Bach *et al.*, 2001; Klock *et al.*, 1998). In the cardiovascular system, one origin of apoptotic cells are infiltrating leuko-cytes (neutrophils) which accumulate in the blood vessels to fight bacteria and, in the absence of infectious agents, undergo apoptosis (von Vietinghoff and Ley, 2008). Removal of dying neutrophils could be a critical issue in atherosclerosis when professional phagocytes (macrophages) and smooth muscle cells (Fries *et al.*, 2005) cannot effectively remove the dead cell material (Schrijvers *et al.*, 2005). CLU upregulation in atherosclerotic blood vessels (Ishikawa *et al.*, 1998) could indicate its supportive function in this process.

In the light of the published data, sCLU may play multiple roles in apoptosis, as a chaperone (Wilson and Easterbrook-Smith, 2000), as a transport/uptake vehicle of cytotoxic material such as the Alzheimer dis-ease-associated amyloid beta peptide (Bell *et al.*, 2007; Yerbury and Wilson, 2009), and as an assistant of phagocytosis when tissue cells are in charge to remove the dead cells early, before macrophages gain access to the tissue and are present in sufficient number. These nonimmune cell types, which also include epithelial and endothelial cells, fibroblasts and smooth muscle cells, may require assisting factors for phagocytosis and react to cell debris by upregulating helper proteins such as sCLU.

Altogether, a clue of the multifunctionality of sCLU seems to lie in its ability to bind to a broad spectrum of proteins, in aggregated or denatured form as well as to exposed hydrophobic domains in native proteins. This is apparently achieved by using distinct domains for interaction with different ligands which include the receptor (Calero *et al.*, 1999; Lakins *et al.*, 2002). The finding that sCLU may bind two ligands simultaneously is in agreement with the model which suggests that it functions as an adaptor which bridges membrane particles and dead cells on one hand and receptors such as megalin and LRP-1 on the other, thereby stimulating the phagocytic process (Fig. 1). In conclusion, sCLU binding and disposal activity may have a dual function, masking critically exposed domains and removing potentially harmful complexes from the extracellular space.

V. sCLU IN IMMUNE MODULATION

An active role of sCLU in the immune system had initially been demon-strated by its activity to prevent membrane attack complex formation (Tschopp and French, 1994). The ability of sCLU to enhance the removal of dead cells may also be considered as immune regulatory since it avoids stimulatory reactions by denatured proteins and cell remnants. Dead cell remnants could provoke an inadequate immune reaction that is directed against cellular antigens (Lu et al., 2008; Munoz et al., 2008; Vioritto et al., 2007) or, in the case of massive immune cell invasion and apoptosis, their inefficient removal may provoke chronic inflammation in diseases affecting the airways, such as cystic fibrosis and bronchiectasis (Lawrence and Gilroy, 2007; Vandivier et al., 2002b). The decision, whether or not during phago- cytosis an immune reaction is initiated, will depend on the type of cell death, apoptosis, or necrosis (Sauter et al., 2000; Savill et al., 2002) and may be further regulated by additional signals. The signaling molecules, which can be components of bacterial cells, products from a viral infection, or material released from dying cells, are recognized by numerous sensor proteins, the most prominent ones being the toll-like receptors (TLRs). Among the TLRs known to this day, TLR4 responds to lipopolysaccharides (LPS), TLR2 binds to several molecules including lipoproteins from bacteria and TLR3 recognizes double-stranded RNA (dsRNA) that may represent viral replica- tion intermediates or RNA released from necrotic cells. In general, TLR activation results in an immune stimulatory status which, in the context of phagocytosis, will favor an immune reaction in contrast to tolerance (for review, see Viorritto et al., 2007).

In VSMCs, we have recently shown that soluble components from necrotic cells induced expression of both sCLU and MCP-1, the major attractant factor for monocytes (Baiersdörfer et al., manuscript submitted for publica- tion). In this system, RNA appears as being the key component producing the response of necrotic material, since the activity was reduced by RNAse treatment. Furthermore, in addition to cellular RNA, poly (I:C), the estab- lished synthetic inducer of TLR3, also stimulated both MCP-1 and sCLU expression (Baiersdörfer et al., manuscript submitted for publication). No induction of the CLU gene upon poly (I:C) treatment was observed in cells not expressing a functional TLR3. The response was restored by ectopic TLR3 expression, clearly demonstrating the role of this receptor in the control of CLU expression.

It is known that the TLR3 agonist dsRNA acts in an immune stimulatory fashion, inducing interferon type I and promoting an immune response. In addition, our results suggest that challenge of cells with dsRNA stimulates the synthesis of cytoprotective clusterin and the chemokine MCP-1 via TLR3. Following MCP-1-dependent attraction of monocytes, the cells will differentiate into macrophages (or dendritic cells) which will become active in dead cell phagocytosis, and possibly in an immune response. CLU might be induced in order to function as an immune suppressive factor. This model is supported by a number of activities of clusterin: inhibition of complement activation (Tschopp and French, 1994); blockade of the neutrophil protease MT6-MMP/MMP-25 (Matsuda et al., 2003); modulation of NF-KappaB signaling, an activity which has been suggested for intracellular CLU

(Deveauchelle *et al.*, 2006; Santilli *et al.*, 2003) thereby modulating a pathway that carries key functions in immune cells (Tan *et al.*, 2005; Yoshimura *et al.*, 2001); suppression of inflammation and secondary immune response which limits myocarditis in wild-type compared to CLU knockout mice (McLaughlin *et al.*, 2000); and, as discussed here, prevention of inflammatory reactions by assisting in the removal of dead cells.

Moreover, the observation that sCLU is cell protective against TNF-α-induced cell death (Ammar and Closset, 2008; Sintich *et al.*, 1999) also indicates that it may function as a suppressor of cellular immune response (Aggarwal, 2003; Chavez-Galan *et al.*, 2009). In the same context, immune suppressive activities are suggested by an association of CLU gene expression with autoimmune diseases. In patients suffering from systemic lupus erythematosus (SLE), a complex autoimmune disease with unknown origin, a correlation of the incidence of the disease with low sCLU serum levels was reported (Newkirk *et al.*, 1999). Furthermore, in SLE an impaired phagocytosis of dead cells by macrophages (Baumann *et al.*, 2002; Herrmann *et al.*, 1998) suggests a correlation of inefficient dead cell removal and autoimmune risk.

The concomitant upregulation of sCLU and the monocyte chemoattractant MCP-1 in VSMCs by necrotic cells or by the TLR3 agonist poly (I:C) points to a role of sCLU in cardiovascular diseases. Along these lines, sCLU has been reported to induce nodule formation when overexpressed in VSMCs or when added extracellularly. Furthermore, sCLU also stimulated VSMC migration (Millis *et al.*, 2001; Moulson and Millis, 1999). This suggests that sCLU triggers the reversible switch of VSMC phenotypes, differentiated or migratory, with the latter status believed to contribute to atherosclerotic diseases when VSMCs migrate to the media and intima of blood vessels (Owens *et al.*, 2004; Schober and Zernecke, 2007).

Atherosclerosis has many features of an immunological disease, such as endothelial cell activation allowing immune cell attachment, deposition of inflammatory markers like C-reactive protein (CRP), or the presence of complement proteins in atherosclerotic plaques (Funk and FitzGerald, 2007; Pereira and Borba, 2008; Schillinger and Minar, 2005). The exact role of sCLU in atherosclerosis is not clear yet, however, the accumulation of sCLU in atherosclerotic lesions (Ishikawa *et al.*, 1998; Witte *et al.*, 1993) and the prevention of lipid oxidation by sCLU (Navab *et al.*, 1997) indicates a protective function.

In atherosclerosis, LDL can be modified by various processes, the most prominent one being oxidation that produces oxLDL, which is thought to promote foam cell formation (Berliner *et al.*, 1995). In addition, an enzymatic modification leading to so-called E-LDL has been described (Bhakdi *et al.*, 1995; Torzewski *et al.*, 2004). E-LDL can also promote foam cell formation and stimulate complement activation (Bhakdi *et al.*, 2004). sCLU may interfere with these processes in two ways, firstly, by acting as a complement inhibitor, and secondly, by direct binding to modified LDL, thereby protecting cells from the cytotoxic effects of the potentially harmful lipoprotein (Schwarz *et al.*, 2008).

VI. ROLE OF sCLU IN AMYLOID BETA (Aβ) CLEARANCE IN M. ALZHEIMER

One of the hallmarks of M. Alzheimer (M.A.) is the massive appearance of extracellular amyloid plaques. The main constituent of these plaques is the amyloid β (Aβ)-peptide, with the predominant forms A$\beta_{(1-40)}$ and A$\beta_{(1-42)}$, which both are generated by proteolytic processing of the APP precursor protein (Selkoe, 2001). Neuronal toxicity of Aβ peptide depends on the amyloid peptide variant and on its self-aggregating property to form oligomeric, prefibrillar assemblies (Bucciantini *et al.*, 2002; Walsh *et al.*, 2002).

The generation of the plaque reflects the accumulation of Aβ which is due to an imbalance between Aβ anabolism and catabolism. While the mechanism of Aβ biogenesis is extensively studied and well characterized, the catabolism of Aβ has only more recently attracted attention. It is suggested that Aβ, secreted from neurons, is degraded by proteases like neprilysin and insulin-degrading enzyme and cleared by astrocytes and microglia via endocytotic/phagocytotic mechanisms (Iwata *et al.*, 2000; Paresce *et al.*, 1996; Qiu *et al.*, 1998; Wyss-Coray *et al.*, 2003). Recent evidence indicates that bidirectional transport across the blood–brain barrier (BBB) plays an important role in the regulation of brain Aβ load. While the influx of Aβ across the BBB into the brain is mediated by direct binding of Aβ to the receptor for advanced glycation end products (RAGE), lipoprotein receptors have been suggested to facilitate the clearance of Aβ from the brain into the blood by transcytosis (Bell *et al.*, 2007; Shibata *et al.*, 2000; Zlokovic, 2004). Remarkably, these lipoprotein receptors do not bind Aβ directly, but only as a complex ligand. LDL-receptor-related protein (LRP-1) binds Aβ complexed to apolipoprotein E (apoE) or α2-macroglobulin (α2M), which both act as carrier proteins and increase the clearance of Aβ (Deane *et al.*, 2008; Kang *et al.*, 2000).

Intriguingly, sCLU also binds to Aβ. This high-affinity interaction (K_d = 2–4 nM) results in the formation of stable complexes with a stoichiometry of 1:1 (Calero *et al.*, 2000; Ghiso *et al.*, 1993). These complexes bind to megalin (LRP-2), which is expressed in vascular CNS tissues including the choroid plexus, the BBB endothelium, and the ependyma (Chun *et al.*, 1999; Hammad *et al.*, 1997). Perfusion studies in guinea pig demonstrated that this receptor is utilized for the efficient transport of sCLU–A$\beta_{(1-40)}$ complexes from the blood across the BBB into the brain (Zlokovic *et al.*, 1996). Although sCLU–A$\beta_{(1-40)}$ complexes have a 2.4-fold higher affinity for megalin than free circulating sCLU, the fact, that plasma levels of free sCLU exceed by more than 5000-fold the level of soluble Aβ suggests that at physiological concentrations this transport mechanism is saturated by free sCLU. Therefore, the sCLU-dependent transport of Aβ via megalin-mediated transcytosis is considered to have no significant influence on the influx of plasma Aβ into the brain (Calero *et al.*, 2000; Shayo *et al.*, 1997). However, data from a recently published study point to a role of this transport route in the efflux of Aβ from the brain. Using an established *in vivo* technique, Bell and coworkers investigated the clearance of Aβ and apolipoproteins administered into mouse brain interstitial fluid (ISF). They demonstrate that sCLU is rapidly eliminated from brain ISF across the BBB via megalin-dependent transcytosis and, furthermore, that binding of highly pathogenic A$\beta_{(1-42)}$ to sCLU accelerates A$\beta_{(1-42)}$ clearance at the BBB by 83% (Bell *et al.*, 2007). The authors suggest that at physiological concentrations the net transport of soluble Aβ via sCLU/megalin at the BBB favors its efflux from the brain.

In addition to modulating Aβ transport, sCLU could also be involved in the local removal of soluble Aβ form the ISF by receptor-mediated endocytosis of sCLU-bound Aβ into brain cells followed by the intracellular degradation of the complexes. In support of this idea, uptake of Aβ into differentiated mouse teratocarcinoma F9 cells upon coincubation with sCLU is enhanced and preformed sCLU–Aβ complexes are internalized and degraded (Hammad *et al.*, 1997). Both processes could be blocked by receptor-associated protein (RAP), an antagonist of all LDL-receptor family members, and by anti-megalin antibodies, demonstrating that sCLU–Aβ complexes are endocytosed via megalin. The participation of other lipoprotein receptors, such as LRP-1 was, however, not analyzed in this study.

Exposition to extracellular Aβ leads to an activation of astrocytes and to a rise in cell-associated sCLU with a concomitant decline in the amount of sCLU in the culture medium pointing to an enhanced endocytosis of sCLU in the presence of Aβ peptide (LaDu *et al.*, 2000). On the basis of the expression of LDL-receptor family members and their different affinities to RAP, the authors suggested that the cellular activation by Aβ is mediated by LRP-1. Internalization of sCLU, however, is not inhibited by RAP, even at high concentrations, indicating the involvement of additional mechanisms in the endocytosis of sCLU–Aβ complexes in astrocytes. Interestingly, induction of cytoplasmic vacuoles containing fibrillar amyloid material in human and rat astrocytes upon exposure to A$\beta_{(1-42)}$ peptides and fibrils is paralleled by an enhanced expression of sCLU (Nuutinen *et al.*, 2007). Furthermore, Aβ clearance from the cerebrospinal fluid by uptake into macrophage-like U937 cells appears to involve extracellular chaperones such as sCLU

(Yerbury and Wilson, 2009). Finally, mice lacking expression of apoE and clusterin (PDAPP, apoE$^{-/-}$, clusterin$^{-/-}$ mice) show both, an early onset and an increase of brain Aβ deposition in an Alzheimer's mouse model (DeMattos *et al.*, 2004).

In conclusion, these results highlight the capacity of sCLU to enhance the clearance of Aβ from the brain, either by accelerated transport across the BBB or by local endocytosis and degradation, and thereby may aid lowering Aβ burden in the brain, which is thought to play a central role in development of Alzheimer's pathology.

In addition to promoting Aβ clearance, the binding of sCLU–Aβ complexes to endocytotic receptors could also affect cellular signaling. As discussed above, interaction of sCLU with megalin/LRP-2 is suggested to elicit prosurvival signaling via the PI3 kinase/Akt pathway (Ammar and Closset, 2008). sCLU also induces the Ras-dependent Raf-1/mitogen-activated protein kinase kinase (MEK/ERK) cascade and a proliferative response in astrocytes via an epidermal growth factor (EGFR)-dependent mechanism (Shim *et al.*, 2009). The latter signaling cascade is not initiated by direct binding of sCLU to EGFR. Given that members of the LDLR family have, in addition to their endocytic function, the capacity to modulate receptor tyrosine kinase signaling via cross-talk mechanisms (Boucher *et al.*, 2002; Newton *et al.*, 2005), it is tempting to speculate that binding to any of these receptors could be involved in the induction of EGFR-dependent proliferation of astrocytes by sCLU. Also, it will be of prime interest to analyze whether binding of sCLU, apoE, or α-2M, either in free form or in complex with Aβ, to the members of the LDL-receptor family triggers prosurvival signaling in neurons and glia cells, which would counteract the detrimental effects of Aβ.

VII. CONCLUSIONS AND PERSPECTIVES

Extracellular clusterin (sCLU) appears as a multifunctional cell protector which sequesters harmful agents during tissue remodeling and degeneration, and helps with their removal. Various disturbances such as hypoxia, stress, cell death, or injury increase sCLU synthesis and thereby stimulate the removal of cell remnants by targeting them to members of the LDLR family, followed by cellular uptake via endocytosis or phagocytosis. Debris removal is impaired in diseases such as atherosclerosis contributing to atherosclerotic plaques, in certain airway diseases with chronic inflammation, and in pathological conditions with inflammation that can lead to autoimmune diseases. In cancer, sCLU is believed to exert its cell-protective function via receptor binding; therefore, an intriguing issue for the future will be the identification of the sCLU receptor(s) in specific tumors and of subsequently induced signal transduction. This will open two options for therapeutic interference with CLU tumor cell protection, prevention of sCLU receptor activation, or blockade of the resulting signaling. In M. Alzheimer, sCLU may have at least two functions modulating Aβ toxicity, one interfering with aggregation and one stimulating Aβ clearance by endocytosis or transcytosis, and possibly by damping the inflammatory response elicited by microglial activation. In all these instances, sCLU could prevent the pathological outcome that may result from harmful extracellular material, by removing toxic substances and potential antigens and by protecting the vital cells in the affected tissue from cell death. This action, which is beneficial in normal physiology, is undesirable in tumors, where it might counteract apoptosis induced by immune cells or by poor vasculation. The role of sCLU, in dead cell removal and cell protection could turn it into being a key player in several pathologies, protective, for example, in autoimmune and cardiovascular diseases and in M. Alzheimer, and life-threatening in tumors.

REFERENCES

Aggarwal, B. B. (2003). Signalling pathways of the TNF superfamily: A double-edged sword. *Nat. Rev. Immunol.* **3**, 745–756.

Ahn, H. J., Bae, J., Lee, S., Ko, J. E., Yoon, S., Kim, S. J., and Sakuragi, N. (2008). Differential expression of clusterin according to histological type of endometrial carcinoma. *Gynecol. Oncol.* **110**, 222–229.

Ammar, H., and Closset, J. L. (2008). Clusterin activates survival through the phosphatidylinositol 3-kinase/Akt pathway. *J. Biol. Chem.* **283**, 12851–12861.

Bach, U. C., Baiersdorfer, M., Klock, G., Cattaruzza, M., Post, A., and Koch-Brandt, C. (2001). Apoptotic cell debris and phosphatidylserine-containing lipid vesicles induce apolipoprotein J (clusterin) gene expression in vital fibroblasts. *Exp. Cell Res.* **265**, 11–20.

Baiersdörfer, M., Schwarz, M., Bergner, O., Heit, A., Wagner, H., Kirschning, C. J., and Koch-Brandt, C. Expression of apolipoprotein J in vascular smooth muscle cells is induced by RNA released from necrotic cells via Toll-like receptor 3. (manuscript submitted for publication).

Bailey, R. W., Aronow, B., Harmony, J. A., and Griswold, M. D. (2002). Heat shock-initiated apoptosis is accelerated and removal of damaged cells is delayed in the testis of clusterin/ApoJ knock-out mice. *Biol. Reprod.* **66**, 1042–1053.

Bajari, T. M., Strasser, V., Nimpf, J., and Schneider, W. J. (2003). A model for modulation of leptin activity by association with clusterin. *FASEB J.* **17**, 1505–1507.

Bandyk, M. G., Sawczuk, I. S., Olsson, C. A., Katz, A. E., and Buttyan, R. (1990). Characterization of the products of a gene expressed during androgen-programmed cell death and their potential use as a marker of urogenital injury. *J. Urol.* **143**, 407–413.

Bartl, M. M., Luckenbach, T., Bergner, O., Ullrich, O., and Koch-Brandt, C. (2001). Multiple receptors mediate apoJ-dependent clearance of cellular debris into nonprofessional phagocytes. *Exp. Cell Res.* **271**, 130–141.

Baumann, I., Kolowos, W., Voll, R. E., Manger, B., Gaipl, U., Neuhuber, W. L., Kirchner, T., Kalden, J. R., and Herrmann, M. (2002). Impaired uptake of apoptotic cells into tingible body macrophages in germinal centers of patients with systemic lupus erythematosus. *Arthritis Rheum.* **46**, 191–201.

Bell, R. D., Sagare, A. P., Friedman, A. E., Bedi, G. S., Holtzman, D. M., Deane, R., and
Zlokovic, B. V. (2007). Transport pathways for clearance of human Alzheimer's amyloid
beta-peptide and apolipoproteins E and J in the mouse central nervous system. *J. Cereb.
Blood Flow Metab.* **27,** 909–918.

Berliner, J. A., Navab, M., Fogelman, A. M., Frank, J. S., Demer, L. L., Edwards, P. A.,
Watson, A. D., and Lusis, A. J. (1995). Atherosclerosis: Basic mechanisms. Oxidation,
inflammation, and genetics. *Circulation* **91,** 2488–2496.

Bhakdi, S., Dorweiler, B., Kirchmann, R., Torzewski, J., Weise, E., Tranum-Jensen, J., Walev, I.,
and Wieland, E. (1995). On the pathogenesis of atherosclerosis: Enzymatic transformation of
human low density lipoprotein to an atherogenic moiety. *J. Exp. Med.* **182,** 1959–1971.

Bhakdi, S., Torzewski, M., Paprotka, K., Schmitt, S., Barsoom, H., Suriyaphol, P., Han, S. R.,
Lackner, K. J., and Husmann, M. (2004). Possible protective role for C-reactive protein in
atherogenesis: Complement activation by modified lipoproteins halts before detrimental
terminal sequence. *Circulation* **109,** 1870–1876.

Boucher, P., Liu, P., Gotthardt, M., Hiesberger, T., Anderson, R. G., and Herz, J. (2002). Platelet-
derived growth factor mediates tyrosine phosphorylation of the cytoplasmic domain of the low
density lipoprotein receptor-related protein in caveolae. *J. Biol. Chem.* **277,** 15507–15513.

Brändström, A., Westin, P., Bergh, A., Cajander, S., and Damber, J. E. (1994). Castration
induces apoptosis in the ventral prostate but not in an androgen-sensitive prostatic adenocarcinoma in the rat. *Cancer Res.* **54,** 3594–3601.

Bucciantini, M., Giannoni, E., Chiti, F., Baroni, F., Formigli, L., Zurdo, J., Taddei, N.,
Ramponi, G., Dobson, C. M., and Stefani, M. (2002). Inherent toxicity of aggregates implies
a common mechanism for protein misfolding diseases. *Nature* **416,** 507–511.

Bursch, W., Gleeson, T., Kleine, L., and Tenniswood, M. (1995). Expression of clusterin
(testosterone-repressed prostate message-2) mRNA during growth and regeneration of rat
liver. *Arch. Toxicol.* **69,** 253–258.

Buttyan, R., Olsson, C. A., Pintar, J., Chang, C., Bandyk, M., Ng, P. Y., and Sawczuk, I. S.
(1989). Induction of the TRPM-2 gene in cells undergoing programmed death. *Mol. Cell.
Biol.* **9,** 3473–3481.

Caccamo, A. E., Scaltriti, M., Caporali, A., D'Arca, D., Scorcioni, F., Candiano, G.,
Mangiola, M., and Bettuzzi, S. (2003). Nuclear translocation of a clusterin isoform is
associated with induction of anoikis in SV40-immortalized human prostate epithelial cells.
Ann. N. Y. Acad. Sci. **1010,** 514–519.

Calero, M., Tokuda, T., Rostagno, A., Kumar, A., Zlokovic, B., Frangione, B., and Ghiso, J.
(1999). Functional and structural properties of lipid-associated apolipoprotein J (clusterin).
Biochem. J. **344**(Pt. 2), 375–383.

Calero, M., Rostagno, A., Matsubara, E., Zlokovic, B., Frangione, B., and Ghiso, J. (2000).
Apolipoprotein J (clusterin) and Alzheimer's disease. *Microsc. Res. Tech.* **50,** 305–315.

Chavez-Galan, L., Arenas-Del Angel, M. C., Zenteno, E., Chavez, R., and Lascurain, R. (2009).
Cell death mechanisms induced by cytotoxic lymphocytes. *Cell. Mol. Immunol* **6,** 15–25.

Chen, H., Tritton, T. R., Kenny, N., Absher, M., and Chiu, J. F. (1996). Tamoxifen induces TGF-
beta 1 activity and apoptosis of human MCF-7 breast cancer cells *in vitro*. *J. Cell. Biochem.*
61, 9–17.

Chen, T., Turner, J., McCarthy, S., Scaltriti, M., Bettuzzi, S., and Yeatman, T. J. (2004).
Clusterin-mediated apoptosis is regulated by adenomatous polyposis coli and is p21
dependent but p53 independent. *Cancer Res.* **64,** 7412–7419.

Chen, S., Bu, G., Takei, Y., Sakamoto, K., Ikematsu, S., Muramatsu, T., and Kadomatsu, K.
(2007). Midkine and LDL-receptor-related protein 1 contribute to the anchorage-independent
cell growth of cancer cells. *J Cell Sci.* **120,** 4009–4015.

Chou, T. Y., Chen, W. C., Lee, A. C., Hung, S. M., Shih, N. Y., and Chen, M. Y. (2009). Clusterin silencing in human lung adenocarcinoma cells induces a mesenchymal-to-epithelial transition through modulating the ERK/Slug pathway. *Cell Signal.* **21**, 704–711.

Chun, J. T., Wang, L., Pasinetti, G. M., Finch, C. E., and Zlokovic, B. V. (1999). Glycoprotein 330/megalin (LRP-2) has low prevalence as mRNA and protein in brain microvessels and choroid plexus. *Exp. Neurol.* **157**, 194–201.

D'Arcangelo, G., Homayouni, R., Keshvara, L., Rice, D. S., Sheldon, M., and Curran, T. (1999). Reelin is a ligand for lipoprotein receptors. *Neuron* **24**, 471–479.

Deane, R., Sagare, A., Hamm, K., Parisi, M., Lane, S., Finn, M. B., Holtzman, D. M., and Zlokovic, B. V. (2008). apoE isoform-specific disruption of amyloid beta peptide clearance from mouse brain. *J. Clin. Invest.* **118**, 4002–4013.

deGoma, E. M., deGoma, R. L., and Rader, D. J. (2008). Beyond high-density lipoprotein cholesterol levels evaluating high-density lipoprotein function as influenced by novel therapeutic approaches. *J. Am. Coll. Cardiol.* **51**, 2199–2211.

DeMattos, R. B., Cirrito, J. R., Parsadanian, M., May, P. C., O'Dell, M. A., Taylor, J. W., Harmony, J. A., Aronow, B. J., Bales, K. R., Paul, S. M., and Holtzman, D. M. (2004). ApoE and clusterin cooperatively suppress Abeta levels and deposition: Evidence that ApoE regulates extracellular Abeta metabolism *in vivo. Neuron* **41**, 193–202.

de Silva, H. V., Stuart, W. D., Duvic, C. R., Wetterau, J. R., Ray, M. J., Ferguson, D. G., Albers, H. W., Smith, W. R., and Harmony, J. A. (1990). A 70-kDa apolipoprotein designated ApoJ is a marker for subclasses of human plasma high density lipoproteins. *J. Biol. Chem.* **265**, 13240–13247.

Devauchelle, V., Essabbani, A., De Pinieux, G., Germain, S., Tourneur, L., Mistou, S., Margottin-Goguet, F., Anract, P., Migaud, H., Le Nen, D., Lequerre, T., Saraux, A., et al. (2006). Characterization and functional consequences of underexpression of clusterin in rheumatoid arthritis. *J. Immunol.* **177**, 6471–6479.

Erwig, L. P., and Henson, P. M. (2008). Clearance of apoptotic cells by phagocytes. *Cell Death Differ.* **15**, 243–250.

Fadok, V. A., Bratton, D. L., Rose, D. M., Pearson, A., Ezekewitz, R. A., and Henson, P. M. (2000). A receptor for phosphatidylserine-specific clearance of apoptotic cells. *Nature* **405**, 85–90.

Farquhar, M. G., Saito, A., Kerjaschki, D., and Orlando, R. A. (1995). The Heymann nephritis antigenic complex: Megalin (gp330) and RAP. *J. Am. Soc. Nephrol.* **6**, 35–47.

French, L. E., Wohlwend, A., Sappino, A. P., Tschopp, J., and Schifferli, J. A. (1994). Human clusterin gene expression is confined to surviving cells during *in vitro* programmed cell death. *J. Clin. Invest.* **93**, 877–884.

Frey, B., Munoz, L. E., Pausch, F., Sieber, R., Franz, S., Brachvogel, B., Poschl, E., Schneider, H., Rodel, F., Sauer, R., Fietkau, R., Herrmann, M., et al. (2008). The immune reaction against allogeneic necrotic cells is reduced in AnnexinA5 knock out mice whose macrophages display an anti-inflammatory phenotype. *J. Cell. Mol. Med.* **20**, 20.

Fries, D. M., Lightfoot, R., Koval, M., and Ischiropoulos, H. (2005). Autologous apoptotic cell engulfment stimulates chemokine secretion by vascular smooth muscle cells. *Am. J. Pathol.* **167**, 345–353.

Funk, C. D., and FitzGerald, G. A. (2007). COX-2 inhibitors and cardiovascular risk. *J. Cardiovasc. Pharmacol.* **50**, 470–479.

Ghiso, J., Matsubara, E., Koudinov, A., Choi-Miura, N. H., Tomita, M., Wisniewski, T., and Frangione, B. (1993). The cerebrospinal-fluid soluble form of Alzheimer's amyloid beta is complexed to SP-40,40 (apolipoprotein J), an inhibitor of the complement membrane-attack complex. *Biochem. J.* **293**(Pt. 1), 27–30.

Gordon, D. J., and Rifkind, B. M. (1989). High-density lipoprotein—The clinical implications of recent studies. *N. Engl. J. Med.* **321**, 1311–1316.

Greijer, A. E., and van der Wall, E. (2004). The role of hypoxia inducible factor 1 (HIF-1) in hypoxia induced apoptosis. *J. Clin. Pathol.* **57**, 1009–1014.

Hammad, S. M., Ranganathan, S., Loukinova, E., Twal, W. O., and Argraves, W. S. (1997). Interaction of apolipoprotein J-amyloid beta-peptide complex with low density lipoprotein receptor-related protein-2/megalin. A mechanism to prevent pathological accumulation of amyloid beta-peptide. *J. Biol. Chem.* **272**, 18644–18649.

Hart, S. P., Smith, J. R., and Dransfield, I. (2004). Phagocytosis of opsonized apoptotic cells: Roles for 'old-fashioned' receptors for antibody and complement. *Clin. Exp. Immunol.* **135**, 181–185.

Henson, P. M., and Hume, D. A. (2006). Apoptotic cell removal in development and tissue homeostasis. *Trends Immunol.* **27**, 244–250.

Herrmann, M., Voll, R. E., Zoller, O. M., Hagenhofer, M., Ponner, B. B., and Kalden, J. R. (1998). Impaired phagocytosis of apoptotic cell material by monocyte-derived macrophages from patients with systemic lupus erythematosus. *Arthritis Rheum.* **41**, 1241–1250.

Hiesberger, T., Trommsdorff, M., Howell, B. W., Goffinet, A., Mumby, M. C., Cooper, J. A., and Herz, J. (1999). Direct binding of Reelin to VLDL receptor and ApoE receptor 2 induces tyrosine phosphorylation of disabled-1 and modulates tau phosphorylation. *Neuron* **24**, 481–489.

Holt, S. K., Karyadi, D. M., Kwon, E. M., Stanford, J. L., Nelson, P. S., and Ostrander, E. A. (2008). Association of megalin genetic polymorphisms with prostate cancer risk and prognosis. *Clin. Cancer Res.* **14**, 3823–3831.

Ishikawa, Y., Akasaka, Y., Ishii, T., Komiyama, K., Masuda, S., Asuwa, N., Choi-Miura, N. H., and Tomita, M. (1998). Distribution and synthesis of apolipoprotein J in the atherosclerotic aorta. *Arterioscler. Thromb. Vasc. Biol.* **18**, 665–672.

Iwata, N., Tsubuki, S., Takaki, Y., Watanabe, K., Sekiguchi, M., Hosoki, E., Kawashima-Morishima, M., Lee, H. J., Hama, E., Sekine-Aizawa, Y., and Saido, T. C. (2000). Identification of the major Abeta1–42-degrading catabolic pathway in brain parenchyma: Suppression leads to biochemical and pathological deposition. *Nat. Med.* **6**, 143–150.

Jeannin, P., Jaillon, S., and Delneste, Y. (2008). Pattern recognition receptors in the immune response against dying cells. *Curr. Opin. Immunol.* **20**, 530–537.

Jenne, D. E., Lowin, B., Peitsch, M. C., Bottcher, A., Schmitz, G., and Tschopp, J. (1991). Clusterin (complement lysis inhibitor) forms a high density lipoprotein complex with apolipoprotein A-I in human plasma. *J. Biol. Chem.* **266**, 11030–11036.

Jo, H., Jia, Y., Subramanian, K. K., Hattori, H., and Luo, H. R. (2008). Cancer cell-derived clusterin modulates the phosphatidylinositol 3¢-kinase-Akt pathway through attenuation of insulin-like growth factor 1 during serum deprivation. *Mol. Cell. Biol.* **28**, 4285–4299.

Józefowski, S., Arredouani, M., Sulahian, T., and Kobzik, L. (2005). Disparate regulation and function of the class A scavenger receptors SR-AI/II and MARCO. *J. Immunol.* **175**, 8032–8041.

Kadomatsu, K., Anzano, M. A., Slayter, M. V., Winokur, T. S., Smith, J. M., and Sporn, M. B. (1993). Expression of sulfated glycoprotein 2 is associated with carcinogenesis induced by N-nitroso-N-methylurea in rat prostate and seminal vesicle. *Cancer Res.* **53**, 1480–1483.

Kang, D. E., Pietrzik, C. U., Baum, L., Chevallier, N., Merriam, D. E., Kounnas, M. Z., Wagner, S. L., Troncoso, J. C., Kawas, C. H., Katzman, R., and Koo, E. H. (2000). Modulation of amyloid beta-protein clearance and Alzheimer's disease susceptibility by the LDL receptor-related protein pathway. *J. Clin. Invest.* **106**, 1159–1166.

Kawamura, A., Baitsch, D., Telgmann, R., Feuerborn, R., Weissen-Plenz, G., Hagedorn, C., Saku, K., Brand-Herrmann, S. M., von Eckardstein, A., Assmann, G., and Nofer, J. R. (2007). Apolipoprotein E interrupts interleukin-1beta signaling in vascular smooth muscle cells. *Arterioscler. Thromb. Vasc. Biol.* **27**, 1610–1617.

Kim, J. H., Yu, Y. S., Kim, K. W., and Min, B. H. (2007). The role of clusterin in *in vitro* ischemia of human retinal endothelial cells. *Curr. Eye Res.* **32**, 693–698.

Klock, G., Storch, S., Rickert, J., Gutacker, C., and Koch-Brandt, C. (1998). Differential regulation of the clusterin gene by Ha-ras and c-myc oncogenes and during apoptosis. *J. Cell. Physiol.* **177,** 593 605.

Koch-Brandt, C., and Morgans, C. (1996). Clusterin: A role in cell survival in the face of apoptosis? *Prog. Mol. Subcell. Biol* **16,** 130–149.

Kolb, S., Vranckx, R., Huisse, M. G., Michel, J. B., and Meilhac, O. (2007). The phosphatidylserine receptor mediates phagocytosis by vascular smooth muscle cells. *J. Pathol.* **212,** 249–259.

Kounnas, M. Z., Loukinova, E. B., Stefansson, S., Harmony, J. A., Brewer, B. H., Strickland, D. K., and Argraves, W. S. (1995). Identification of glycoprotein 330 as an endocytic receptor for apolipoprotein J/clusterin. *J. Biol. Chem.* **270,** 13070–13075.

Kujiraoka, T., Hattori, H., Miwa, Y., Ishihara, M., Ueno, T., Ishii, J., Tsuji, M., Iwasaki, T., Sasaguri, Y., Fujioka, T., Saito, S., Tsushima, M., *et al.* (2006). Serum apolipoprotein J in health, coronary heart disease and type 2 diabetes mellitus. *J. Atheroscler. Thromb.* **13,** 314–322.

Kyprianou, N., English, H. F., Davidson, N. E., and Isaacs, J. T. (1991). Programmed cell death during regression of the MCF-7 human breast cancer following estrogen ablation. *Cancer Res.* **51,** 162–166.

LaDu, M. J., Shah, J. A., Reardon, C. A., Getz, G. S., Bu, G., Hu, J., Guo, L., and van Eldik, L. J. (2000). Apolipoprotein E receptors mediate the effects of beta-amyloid on astrocyte cultures. *J. Biol. Chem.* **275,** 33974–33980.

Lakins, J. N., Poon, S., Easterbrook-Smith, S. B., Carver, J. A., Tenniswood, M. P., and Wilson, M. R. (2002). Evidence that clusterin has discrete chaperone and ligand binding sites. *Biochemistry* **41,** 282–291.

Lawrence, T., and Gilroy, D. W. (2007). Chronic inflammation: A failure of resolution? *Int. J. Exp. Pathol.* **88,** 85–94.

Leskov, K. S., Klokov, D. Y., Li, J., Kinsella, T. J., and Boothman, D. A. (2003). Synthesis and functional analyses of nuclear clusterin, a cell death protein. *J. Biol. Chem.* **278,** 11590–11600.

Lillis, A. P., Greenlee, M. C., Mikhailenko, I., Pizzo, S. V., Tenner, A. J., Strickland, D. K., and Bohlson, S. S. (2008). Murine low-density lipoprotein receptor-related protein 1 (LRP) is required for phagocytosis of targets bearing LRP ligands but is not required for C1q-triggered enhancement of phagocytosis. *J. Immunol.* **181,** 364–373.

Lu, J. H., Teh, B. K., Wang, L., Wang, Y. N., Tan, Y. S., Lai, M. C., and Reid, K. B. (2008). The classical and regulatory functions of C1q in immunity and autoimmunity. *Cell. Mol. Immunol.* **5,** 9–21.

Marzolo, M. P., Yuseff, M. I., Retamal, C., Donoso, M., Ezquer, F., Farfan, P., Li, Y., and Bu, G. (2003). Differential distribution of low-density lipoprotein-receptor-related protein (LRP) and megalin in polarized epithelial cells is determined by their cytoplasmic domains. *Traffic* **4,** 273–288.

Matsuda, A., Itoh, Y., Koshikawa, N., Akizawa, T., Yana, I., and Seiki, M. (2003). Clusterin, an abundant serum factor, is a possible negative regulator of MT6-MMP/MMP-25 produced by neutrophils. *J. Biol. Chem.* **278,** 36350–36357.

May, P., Herz, J., and Bock, H. H. (2005). Molecular mechanisms of lipoprotein receptor signalling. *Cell. Mol. Life Sci.* **62,** 2325–2338.

McLoughlin, D. M., and Miller, C. C. (2008). The FE65 proteins and Alzheimer's disease. *J. Neurosci. Res.* **86,** 744–754.

McLaughlin, L., Zhu, G., Mistry, M., Ley-Ebert, C., Stuart, W. D., Florio, C. J., Groen, P. A., Witt, S. A., Kimball, T. R., Witte, D. P., Harmony, J. A., and Aronow, B. J. (2000). Apolipoprotein J/clusterin limits the severity of murine autoimmune myocarditis. *J. Clin. Invest.* **106,** 1105–1113.

Millis, A. J., Luciani, M., McCue, H. M., Rosenberg, M. E., and Moulson, C. L. (2001). Clusterin regulates vascular smooth muscle cell nodule formation and migration. *J. Cell. Physiol.* **186**, 210–219.

Mitchell, J. E., Cvetanovic, M., Tibrewal, N., Patel, V., Colamonici, O. R., Li, M. O., Flavell, R. A., Levine, J. S., Birge, R. B., and Ucker, D. S. (2006). The presumptive phosphatidylserine receptor is dispensable for innate anti-inflammatory recognition and clearance of apoptotic cells. *J. Biol. Chem.* **281**, 5718–5725.

Miyake, H., Nelson, C., Rennie, P. S., and Gleave, M. E. (2000). Acquisition of chemoresistant phenotype by overexpression of the antiapoptotic gene testosterone-repressed prostate message-2 in prostate cancer xenograft models. *Cancer Res.* **60**, 2547–2554.

Miyanishi, M., Tada, K., Koike, M., Uchiyama, Y., Kitamura, T., and Nagata, S. (2007). Identification of Tim4 as a phosphatidylserine receptor. *Nature* **450**, 435–439.

Moretti, R. M., Marelli, M. M., Mai, S., Cariboni, A., Scaltriti, M., Bettuzzi, S., and Limonta, P. (2007). Clusterin isoforms differentially affect growth and motility of prostate cells: Possible implications in prostate tumorigenesis. *Cancer Res.* **67**, 10325–10333.

Moulson, C. L., and Millis, A. J. (1999). Clusterin (Apo J) regulates vascular smooth muscle cell differentiation *in vitro*. *J. Cell. Physiol.* **180**, 355–364.

Munoz, L. E., van Bavel, C., Franz, S., Berden, J., Herrmann, M., and van der Vlag, J. (2008). Apoptosis in the pathogenesis of systemic lupus erythematosus. *Lupus* **17**, 371–375.

Navab, M., Imes, S. S., Hama, S. Y., Hough, G. P., Ross, L. A., Bork, R. W., Valente, A. J., Berliner, J. A., Drinkwater, D. C., Laks, H., and Fogelman, A. M. (1991). Monocyte transmigration induced by modification of low density lipoprotein in cocultures of human aortic wall cells is due to induction of monocyte chemotactic protein 1 synthesis and is abolished by high density lipoprotein. *J. Clin. Invest.* **88**, 2039–2046.

Navab, M., Hama-Levy, S., Van Lenten, B. J., Fonarow, G. C., Cardinez, C. J., Castellani, L. W., Brennan, M. L., Lusis, A. J., Fogelman, A. M., and La Du, B. N. (1997). Mildly oxidized LDL induces an increased apolipoprotein J/paraoxonase ratio. *J. Clin. Invest.* **99**, 2005–2019.

Navab, M., Anantharamaiah, G. M., Reddy, S. T., Van Lenten, B. J., Wagner, A. C., Hama, S., Hough, G., Bachini, E., Garber, D. W., Mishra, V. K., Palgunachari, M. N., and Fogelman, A. M. (2005). An oral apoJ peptide renders HDL antiinflammatory in mice and monkeys and dramatically reduces atherosclerosis in apolipoprotein E-null mice. *Arterioscler. Thromb. Vasc. Biol.* **25**, 1932–1937.

Navab, M., Anantharamaiah, G. M., Reddy, S. T., Van Lenten, B. J., Buga, G. M., and Fogelman, A. M. (2007). Peptide mimetics of apolipoproteins improve HDL function. *J. Clin. Lipidol.* **1**, 142–147.

Newkirk, M. M., Apostolakos, P., Neville, C., and Fortin, P. R. (1999). Systemic lupus erythematosus, a disease associated with low levels of clusterin/apoJ, an antiinflammatory protein. *J. Rheumatol.* **26**, 597–603.

Newton, C. S., Loukinova, E., Mikhailenko, I., Ranganathan, S., Gao, Y., Haudenschild, C., and Strickland, D. K. (2005). Platelet-derived growth factor receptor-beta (PDGFR-beta) activation promotes its association with the low density lipoprotein receptor-related protein (LRP). Evidence for co-receptor function. *J. Biol. Chem.* **280**, 27872–27878.

Nuutinen, T., Huuskonen, J., Suuronen, T., Ojala, J., Miettinen, R., and Salminen, A. (2007). Amyloid-beta 1–42 induced endocytosis and clusterin/apoJ protein accumulation in cultured human astrocytes. *Neurochem. Int.* **50**, 540–547.

Owens, G. K., Kumar, M. S., and Wamhoff, B. R. (2004). Molecular regulation of vascular smooth muscle cell differentiation in development and disease. *Physiol. Rev.* **84**, 767–801.

Parczyk, K., Pilarsky, C., Rachel, U., and Koch-Brandt, C. (1994). Gp80 (clusterin; TRPM-2) mRNA level is enhanced in human renal clear cell carcinomas. *J. Cancer Res. Clin. Oncol.* **120**, 186–188.

Paresce, D. M., Ghosh, R. N., and Maxfield, F. R. (1996). Microglial cells internalize aggregates of the Alzheimer's disease amyloid beta-protein via a scavenger receptor. *Neuron* **17**, 553–565.

Park, D., Tosello-Trampont, A. C., Elliott, M. R., Lu, M., Haney, L. B., Ma, Z., Klibanov, A. L., Mandell, J. W., and Ravichandran, K. S. (2007). BAI1 is an engulfment receptor for apoptotic cells upstream of the ELMO/Dock180/Rac module. *Nature* **450**, 430–434.

Patel, V. A., Longacre, A., Hsiao, K., Fan, H., Meng, F., Mitchell, J. E., Rauch, J., Ucker, D. S., and Levine, J. S. (2006). Apoptotic cells, at all stages of the death process, trigger characteristic signaling events that are divergent from and dominant over those triggered by necrotic cells: Implications for the delayed clearance model of autoimmunity. *J. Biol. Chem.* **281**, 4663–4670.

Pereira, I. A., and Borba, E. F. (2008). The role of inflammation, humoral and cell mediated autoimmunity in the pathogenesis of atherosclerosis. *Swiss Med. Wkly* **138**, 534–539.

Platt, N., da Silva, R. P., and Gordon, S. (1998). Recognizing death: The phagocytosis of apoptotic cells. *Trends Cell Biol.* **8**, 365–372.

Poulios, E., Trougakos, I. P., and Gonos, E. S. (2006). Comparative effects of hypoxia on normal and immortalized human diploid fibroblasts. *Anticancer Res.* **26**, 2165–2168.

Qiu, W. Q., Walsh, D. M., Ye, Z., Vekrellis, K., Zhang, J., Podlisny, M. B., Rosner, M. R., Safavi, A., Hersh, L. B., and Selkoe, D. J. (1998). Insulin-degrading enzyme regulates extracellular levels of amyloid beta-protein by degradation. *J. Biol. Chem.* **273**, 32730–32738.

Ranney, M. K., Ahmed, I. S., Potts, K. R., and Craven, R. J. (2007). Multiple pathways regulating the anti-apoptotic protein clusterin in breast cancer. *Biochim. Biophys. Acta* **1772**, 1103–1111.

Redondo, M., Villar, E., Torres-Munoz, J., Tellez, T., Morell, M., and Petito, C. K. (2000). Overexpression of clusterin in human breast carcinoma. *Am. J. Pathol.* **157**, 393–399.

Santilli, G., Aronow, B. J., and Sala, A. (2003). Essential requirement of apolipoprotein J (clusterin) signaling for IkappaB expression and regulation of NF-kappaB activity. *J. Biol. Chem.* **278**, 38214–38219.

Sauter, B., Albert, M. L., Francisco, L., Larsson, M., Somersan, S., and Bhardwaj, N. (2000). Consequences of cell death: Exposure to necrotic tumor cells, but not primary tissue cells or apoptotic cells, induces the maturation of immunostimulatory dendritic cells. *J. Exp. Med.* **191**, 423–434.

Savill, J., and Gregory, C. (2007). Apoptotic PS to phagocyte TIM-4: Eat me. *Immunity* **27**, 830–832.

Savill, J., Dransfield, I., Gregory, C., and Haslett, C. (2002). A blast from the past: Clearance of apoptotic cells regulates immune responses. *Nat. Rev. Immunol.* **2**, 965–975.

Schillinger, M., and Minar, E. (2005). Restenosis after percutaneous angioplasty: The role of vascular inflammation. *Vasc. Health Risk Manag.* **1**, 73–78.

Schober, A., and Zernecke, A. (2007). Chemokines in vascular remodeling. *Thromb. Haemost.* **97**, 730–737.

Schrijvers, D. M., De Meyer, G. R., Kockx, M. M., Herman, A. G., and Martinet, W. (2005). Phagocytosis of apoptotic cells by macrophages is impaired in atherosclerosis. *Arterioscler. Thromb. Vasc. Biol.* **25**, 1256–1261.

Schwarz, M., Spath, L., Lux, C. A., Paprotka, K., Torzewski, M., Dersch, K., Koch-Brandt, C., Husmann, M., and Bhakdi, S. (2008). Potential protective role of apoprotein J (clusterin) in atherogenesis: Binding to enzymatically modified low-density lipoprotein reduces fatty acid-mediated cytotoxicity. *Thromb. Haemost.* **100**, 110–118.

Selkoe, D. J. (2001). Alzheimer's disease: Genes, proteins, and therapy. *Physiol. Rev.* **81**, 741–766.

Sensibar, J. A., Sutkowski, D. M., Raffo, A., Buttyan, R., Griswold, M. D., Sylvester, S. R., Kozlowski, J. M., and Lee, C. (1995). Prevention of cell death induced by tumor necrosis factor alpha in LNCaP cells by overexpression of sulfated glycoprotein-2 (clusterin). *Cancer Res.* **55**, 2431–2437.

Shannan, B., Seifert, M., Leskov, K., Willis, J., Boothman, D., Tilgen, W., and Reichrath, J. (2006). Challenge and promise: Roles for clusterin in pathogenesis, progression and therapy of cancer. *Cell Death Differ.* **13**, 12–19.

Shayo, M., McLay, R. N., Kastin, A. J., and Banks, W. A. (1997). The putative blood–brain barrier transporter for the beta-amyloid binding protein apolipoprotein j is saturated at physiological concentrations. *Life Sci.* **60**, PL115–PL118.

Shibata, M., Yamada, S., Kumar, S. R., Calero, M., Bading, J., Frangione, B., Holtzman, D. M., Miller, C. A., Strickland, D. K., Ghiso, J., and Zlokovic, B. V. (2000). Clearance of Alzheimer's amyloid-ss(1–40) peptide from brain by LDL receptor-related protein-1 at the blood–brain barrier. *J. Clin. Invest.* **106**, 1489–1499.

Shim, Y. J., Shin, Y. J., Jeong, S. Y., Kang, S. W., Kim, B. M., Park, I. S., and Min, B. H. (2009). Epidermal growth factor receptor is involved in clusterin-induced astrocyte proliferation. *Neuroreport* **20**, 435–439.

Sintich, S. M., Steinberg, J., Kozlowski, J. M., Lee, C., Pruden, S., Sayeed, S., and Sensibar, J. A. (1999). Cytotoxic sensitivity to tumor necrosis factor-alpha in PC3 and LNCaP prostatic cancer cells is regulated by extracellular levels of SGP-2 (clusterin). *Prostate* **39**, 87–93.

Strickland, D. K., and Ranganathan, S. (2003). Diverse role of LDL receptor-related protein in the clearance of proteases and in signaling. *J. Thromb. Haemost.* **1**, 1663–1670.

Strickland, D. K., Kounnas, M. Z., and Argraves, W. S. (1995). LDL receptor-related protein: A multiligand receptor for lipoprotein and proteinase catabolism. *FASEB J.* **9**, 890–898.

Tan, P. H., Sagoo, P., Chan, C., Yates, J. B., Campbell, J., Beutelspacher, S. C., Foxwell, B. M., Lombardi, G., and George, A. J. (2005). Inhibition of NF-kappa B and oxidative pathways in human dendritic cells by antioxidative vitamins generates regulatory T cells. *J. Immunol.* **174**, 7633–7644.

Terrand, J., Bruban, V., Zhou, L., Gong, W., El Asmar, Z., May, P., Zurhove, K., Haffner, P., Philippe, C., Woldt, E., Matz, R. L., Gracia, C., *et al.* (2009). LRP1 controls intracellular cholesterol storage and fatty acid synthesis through modulation of Wnt signaling. *J. Biol. Chem.* **284**, 381–388.

Torzewski, M., Suriyaphol, P., Paprotka, K., Spath, L., Ochsenhirt, V., Schmitt, A., Han, S. R., Husmann, M., Gerl, V. B., Bhakdi, S., and Lackner, K. J. (2004). Enzymatic modification of low-density lipoprotein in the arterial wall: A new role for plasmin and matrix metalloproteinases in atherogenesis. *Arterioscler. Thromb. Vasc. Biol.* **24**, 2130–2136.

Trommsdorff, M., Gotthardt, M., Hiesberger, T., Shelton, J., Stockinger, W., Nimpf, J., Hammer, R. E., Richardson, J. A., and Herz, J. (1999). Reeler/Disabled-like disruption of neuronal migration in knockout mice lacking the VLDL receptor and ApoE receptor 2. *Cell* **97**, 689–701.

Trougakos, I. P., and Gonos, E. S. (2002). Clusterin/apolipoprotein J in human aging and cancer. *Int. J. Biochem. Cell. Biol.* **34**, 1430–1448.

Trougakos, I. P., Lourda, M., Antonelou, M. H., Kletsas, D., Gorgoulis, V. G., Papassideri, I. S., Zou, Y., Margaritis, L. H., Boothman, D. A., and Gonos, E. S. (2009). Intracellular clusterin inhibits mitochondrial apoptosis by suppressing p53-activating stress signals and stabilizing the cytosolic Ku70-Bax protein complex. *Clin. Cancer Res.* **15**, 48–59.

Tschopp, J., and French, L. E. (1994). Clusterin: Modulation of complement function. *Clin. Exp. Immunol.* **97**, 11–14.

Vandivier, R. W., Ogden, C. A., Fadok, V. A., Hoffmann, P. R., Brown, K. K., Botto, M., Walport, M. J., Fisher, J. H., Henson, P. M., and Greene, K. E. (2002a). Role of surfactant proteins A, D, and C1q in the clearance of apoptotic cells *in vivo* and *in vitro*: Calreticulin and CD91 as a common collectin receptor complex. *J. Immunol.* **169**, 3978–3986.

Vandivier, R. W., Fadok, V. A., Hoffmann, P. R., Bratton, D. L., Penvari, C., Brown, K. K., Brain, J. D., Accurso, F. J., and Henson, P. M. (2002b). Elastase-mediated phosphatidylserine receptor cleavage impairs apoptotic cell clearance in cystic fibrosis and bronchiectasis. *J. Clin. Invest.* **109**, 661–670.

Van Lenten, B. J., Hama, S. Y., de Beer, F. C., Stafforini, D. M., McIntyre, T. M., Prescott, S. M., La Du, B. N., Fogelman, A. M., and Navab, M. (1995). Anti-inflammatory HDL becomes pro-inflammatory during the acute phase response. Loss of protective effect of HDL against LDL oxidation in aortic wall cell cocultures. *J. Clin. Invest.* **96**, 2758–2767.

Viorritto, I. C., Nikolov, N. P., and Siegel, R. M. (2007). Autoimmunity versus tolerance: Can dying cells tip the balance? *Clin. Immunol.* **122**, 125–134.

von Vietinghoff, S., and Ley, K. (2008). Homeostatic regulation of blood neutrophil counts. *J. Immunol.* **181**, 5183–5188.

Walsh, D. M., Klyubin, I., Fadeeva, J. V., Cullen, W. K., Anwyl, R., Wolfe, M. S., Rowan, M. J., and Selkoe, D. J. (2002). Naturally secreted oligomers of amyloid beta protein potently inhibit hippocampal long-term potentiation *in vivo*. *Nature* **416**, 535–539.

Watson, A. D., Berliner, J. A., Hama, S. Y., La Du, B. N., Faull, K. F., Fogelman, A. M., and Navab, M. (1995). Protective effect of high density lipoprotein associated paraoxonase. Inhibition of the biological activity of minimally oxidized low density lipoprotein. *J. Clin. Invest.* **96**, 2882–2891.

Wellmann, A., Thieblemont, C., Pittaluga, S., Sakai, A., Jaffe, E. S., Siebert, P., and Raffeld, M. (2000). Detection of differentially expressed genes in lymphomas using cDNA arrays: Identification of clusterin as a new diagnostic marker for anaplastic large-cell lymphomas. *Blood* **96**, 398–404.

Wilson, M. R., and Easterbrook-Smith, S. B. (2000). Clusterin is a secreted mammalian chaperone. *Trends Biochem. Sci.* **25**, 95–98.

Witte, D. P., Aronow, B. J., Stauderman, M. L., Stuart, W. D., Clay, M. A., Gruppo, R. A., Jenkins, S. H., and Harmony, J. A. (1993). Platelet activation releases megakaryocyte-synthesized apolipoprotein J, a highly abundant protein in atheromatous lesions. *Am. J. Pathol.* **143**, 763–773.

Wyss-Coray, T., Loike, J. D., Brionne, T. C., Lu, E., Anankov, R., Yan, F., Silverstein, S. C., and Husemann, J. (2003). Adult mouse astrocytes degrade amyloid-beta *in vitro* and *in situ*. *Nat. Med.* **9**, 453–457.

Yerbury, J. J., and Wilson, M. R. (2009). Extracellular chaperones modulate the effects of Alzheimer's patient cerebrospinal fluid on Abeta(1–42) toxicity and uptake. *Cell Stress Chaperones* [Epub ahead of print], PMID: 19472074.

Yoshimura, S., Bondeson, J., Brennan, F. M., Foxwell, B. M., and Feldmann, M. (2001). Role of NFkappaB in antigen presentation and development of regulatory T cells elucidated by treatment of dendritic cells with the proteasome inhibitor PSI. *Eur. J. Immunol.* **31**, 1883–1893.

You, K. H., Ji, Y. M., and Kwon, O. Y. (2003). Clusterin overexpression is responsible for the anti-apoptosis effect in a mouse neuroblastoma cell line, B103. *Z. Naturforsch. [C]* **58**, 148–151.

Zhang, H., Kim, J. K., Edwards, C. A., Xu, Z., Taichman, R., and Wang, C. Y. (2005). Clusterin inhibits apoptosis by interacting with activated Bax. *Nat. Cell Biol.* **7**, 909–915.

Zhou, J., Schmid, T., Schnitzer, S., and Brune, B. (2006). Tumor hypoxia and cancer progression. *Cancer Lett.* **237**, 10–21.

Zlokovic, B. V. (2004). Clearing amyloid through the blood–brain barrier. *J. Neurochem.* **89**, 807–811.

Zlokovic, B. V., Martel, C. L., Matsubara, E., McComb, J. G., Zheng, G., McCluskey, R. T., Frangione, B., and Ghiso, J. (1996). Glycoprotein 330/megalin: Probable role in receptor-mediated transport of apolipoprotein J alone and in a complex with Alzheimer disease amyloid beta at the blood–brain and blood–cerebrospinal fluid barriers. *Proc. Natl. Acad. Sci. USA* **93**, 4229–4234.

Züllig, S., and Hengartner, M. O. (2004). Cell biology. Tickling macrophages, a serious business. *Science* **304**, 1123–1124.

Zurhove, K., Nakajima, C., Herz, J., Bock, H. H., and May, P. (2008). Gamma-secretase limits the inflammatory response through the processing of LRP1. *Sci. Signal.* **1**, ra15.

Clusterin: A Multifacet Protein at the Crossroad of Inflammation and Autoimmunity

Géraldine Falgarone[*,†] and Gilles Chiocchia[‡,§,¶]

*EA 4222, University of Paris 13, Bobigny, France
†Rheumatology Department, AP-HP, Hôpital Avicenne, Bobigny F-93009, France
‡Institut Cochin, Université Paris Descartes, CNRS (UMR 8104), Paris F-75014, France
§Département d'Immunologie, Inserm, U567, Paris F-75014, France
¶AP-HP, Hôpital Ambroise Paré, Service de Rhumatologie, Boulogne-Billancourt F-92000, France

For years, clusterin has been recognized as a secreted protein and a large number of works demonstrated that this ubiquitously expressed protein has multiple activities. Among the described activities several were related to inflammation and immunity such as its regulatory activity on complement. Then it became clear that a nuclear form of the protein with proapoptotic property existed and more recently that a

Advances in CANCER RESEARCH
0065-230X/09 $35.00
DOI: 10.1016/S0065-230X(09)04008-1

cytoplasmic form could regulate NF-κB pathway. Again, these activities have a strong repercussion in inflammation and immunity. On the other hand, data available on the exact role of CLU in these processes and autoimmunity were quite scarce until recently. Indeed, in the last few years, a differential CLU expression in subtype of T cells, the regulation of CLU expression by proinflammatory cytokines and molecules, the regulation of expression and function of CLU depending on its subcellular localization, the interaction of CLU with nuclear and intracellular proteins were all reported. Adding these new roles of CLU to the already reported functions of this protein allows a better understanding of its role and potential involvement in several inflammatory and immunological processes and, in particular, autoimmunity. In this sense, rheumatoid arthritis appears to be a very attractive disease to build a new paradigm of the role and function of CLU because it makes the link between proliferation, inflammation, and autoimmunity.

We will try to see in this review how to bring altogether the old and new knowledge on CLU with inflammation and autoimmunity. Nevertheless, it is clear that CLU has not yet revealed all its secrets in inflammation and autoimmunty. © 2009 Elsevier Inc.

ABBREVIATIONS

ASN, antisense nucleotide; Bax, BCL2-associated X protein; CCR4, CXCR3, chemiokine receptors; CIC or IC, circulating immune complexes or Immune complexes; CLU, Clusterin; DAMPs, Damage associated molecular pattern; FcgammaR, Fraction constant gamma Receptor; FLS, Fibroblast like synoviocyte; HUVEC, Human Umbilical Vein Endothelial Cells; IL, Interleukin; LPS, lipopolysaccharide; MAC, Membran Attack complex; MHC, Major Histocompatibility Complex; NLR, Nod Like receptor; OA, osteoarthritis; PAMPs, Pathogen associated molecular pattern; RA, rheumatoid arthritis; RLR, Rig-like receptor; SLE, systemic lupus erythematosus; TCC, Terminal complement complex; TH1, 2, 17, T helper lymphocyte type 1, 2, 17; TLR, Toll-like receptor; Treg, Regulatory T cell.

I. INTRODUCTION

Autoimmune diseases are chronic syndromes characterized by relapsing clinical symptoms linked to autoantibodies production and autoreactive T cells arising against autoantigen. Autoantibodies are, in this context, the hallmark of humoral autoimmunity, and autoreactive T cells are the expression of cellular response. These reactions can concern autoantigen-expressing tissues and induce autoimmune diabetes, rheumatoid arthritis (RA), as well as myasthenia gravis and multiple sclerosis. These autoimmune diseases are known to be linked to MCH susceptibly and other genetic conditions, but are, however, subjected to multiple environmental factors. All organs have been described as targets of autoimmunity, but there is some specificity of clinical syndrome related to the different mechanisms relaying the disease.

Infections could be the starting event; disruption of local defense inducing innate immunity responses, destruction of the organs inducing neoantigens formation, as well as the lack of rapid resolution with inflammation persistence can favor the emerging of autoimmunity. The cause and consequence of these phenomena are often linked and studied together.

Even though infections could be the "primum movens" in many autoimmune diseases, genetic background is to be considered too. The next step is the central or peripheral tolerance rupture that ends with autoantibodies or T cell reaction against autoantigen.

II. INFLAMMATION, THE DANGER SIGNAL, AND AUTOIMMUNITY

New theories emerged over the last decade to explain the triggering of immune reactions. Among these theories, one received a lot of development in autoimmunity and cancer immunology, the danger theory (Janeway *et al.*, 1996; Matzinger, 1994). The theory explains that "the immune system does not care about self and nonself, that its primary driving force is the need to detect and protect against danger, and that it does not do the job alone, but receives positive and negative communications from an extended network of other bodily tissues." These signals include bacterial and viral components, called PAMP for pathogen-associated molecular pattern recognized by a family of receptors, the Toll-like receptors. These receptors are expressed by innate immunity cells, mainly phagocytic cells such as macrophages, dendritic cells, polymorphonuclear cells, which all have in common the production or secretion of inflammatory cytokines when stimulated. However, pathogen components are not the only stimulators of these reactions. Indeed, damaged cells themselves can turn out to be the source of danger signals, identified as DAMP for Danger-Associated Molecular Pattern. Stressed cells can be directly recognized by TLR, as well as NLR (Nod-Like Receptor), that mainly recognized bacterial components, and by RLR (Rig-Like Receptor) that recognized viral components. Heat-shock proteins are an important source of DAMP.

These signals are the source of the inflammatory reaction that rise to delete the source of PAMP or DAMP production, relying on several mediators among them complement, nitric oxide radicals, opsonins, leucotrienes, prostaglandins, and proinflammatory cytokines. The aim is to delete the pathological signal by mounting an inflammatory response that should result in deleting of the stress, as well as repairing mechanisms. This second part is linked to proteins that could protect stressed cells (e.g., chaperones), but could balance reparation and regulation mechanisms.

This statement is also one of those described in cancer: how to destroy undesirable cells or tissues, without destroying the entire environment but still controlling "troop army"?

III. IMMUNE SYSTEM HOMEOSTASIS AND REGULATION: THE COMMON PURPOSE OF AUTOIMMUNITY AND CANCER

The immune system has an extraordinary set of mechanisms working together to maintain self-tolerance and homeostasis. Several actors of different frameworks take place to ensure self-tolerance among them apoptosis, anergy, T regulatory cell (Treg) expansion, cytokine production or deviation. Lack of apoptosis of activated T cells, absence of Treg, and inappropriate cytokine balance of Th-1, Th-2, and Th-17 lymphocytes subtypes, can favor the loss of peripheral tolerance, against self-tissue, and/or prevent the production of cytotoxic T cells directed against the tumor (Salmond *et al.*, 2009; Vignali *et al.*, 2008; Wells, 2009). The absolute requirement of homeostasis together with the control of both the inflammation and autoimmunity are sine qua non conditions for treating autoimmune diseases and cancers.

IV. INFLAMMATION

Inflammation is a complex series of cellular and soluble events that occur in response to a danger which could either be a pathogen invasion or several injuries such as crystal deposits, wounds, UV, etc. One of its characteristics is that it does not occur in a specific manner. Indeed, it may occur during the first stage of innate immunity or during a specific immune response, in reaction to cytokine production, for example. It will then perform some antigen clearance, among other goals. Inflammation occurs during the first, second, and third stages, when recovery of homeostasis should take place but does not, because of chronic stimulation, persistence of pathogen, or cellular abnormalities, for example. If the first step of chronic inflammation is not always clearly definite, it is overwhelmingly admitted that chronic inflammation is a self-renewal process.

Damaged tissues release molecules which attract leucocytes on the inflammation site; these leucocytes circulate and cross blood vessel walls to enter the inflamed tissues. For instance, a chemotactic gradient facilitates the migration of neutrophils, macrophages; which for one major function is the phagocytosis of causal molecule or pathogens. At this step, complement is directly involved either in phagocytosis or in chemoattraction.

A. CLU: A Complement Inhibitor and an Acute-Phase Protein

Complement is a collection of about 30 serum and membrane proteins, directly implicated in host defense, but also in autoimmunity pathogenesis. It is a complex system of functional related proteins that are sequentially activated in a tightly controlled cascade. As a part of lysis pathogen and opsonization of pathogen, complement is involved in clearance of immune complexes, enhancement of antigen presentation, enhancement of B cell activation, generation of peptides that are implicated in vasodilatation, phagocyte adhesion and chemotaxis, regulation of complement action itself, and signaling immune response.

Enzyme inhibitors are present at each step of the cascade to tightly regulate the complete system.

Among those complement inhibitors, soluble Clusterin (sCLU) has been described as a link to the complex of membrane attack (C5b9) which is the terminal molecular component which intervenes to disrupt the membrane of pathogen or target cells (Murphy *et al.*, 1988). sCLU was present at moderate levels (35–105 μm/ml) in normal human serum. First observations of sCLU showed that it was deposited in the renal glomeruli of patients with glomerulonephritis but was not found in normal glomeruli. The tissue distribution of sCLU was considered to be close to that of terminal complement components. Furthermore, as it was observed in the membrane complex attack (MAC), it was initially considered to be an additional component complex. sCLU is indeed a part of the complex but its action is to protect cells or tissues from MAC and renders it cytolytically inactive. Others have observed that sCLU is not always linked to MAC, especially in tissue necrosis induced by pulsed laser (French *et al.*, 1992). sCLU and vibronectin bind together to the nascent C5b9 complex, thus rendering it water soluble and lytically inactive. In fact, sCLU specifically binds to C7, the β-subunit of C8 and C9 (Tschopp *et al.*, 1993). The conformational change occurring during the hydrophilic–amphiphilic transition of C9, in the formation of the MAC, exposes the interaction site for sCLU that then exerts its inhibitory effect by interacting with a structural motif common to C7, C8α, and C9b. We will discuss later on the pathophysiology specially implicated in lupus nephritis. The complement regulatory (CR) proteins CLU and vitronectin bind to the MAC and thus prevent cytolysis

(Choi *et al.*, 1989). Both proteins on the MAC link to circulating immune complexes (CIC) in systemic lupus erythematosus (SLE) patients (Chauhan and Moore, 2006). The amount of both CLU and vitronectin associated with MAC–CIC is two- to threefold higher in comparison with those observed in the soluble form of MAC. Moreover, patients with high levels of sCLU and vitronectin demonstrated renal involvement. This could be linked to tissue damage implication of MAC–CIC, or linked to the faculty of sCLU to protect a variety of secretory, mucosal, and other barrier cells from surface-active components of the extracellular environment (Aronow *et al.*, 1993). For example, sCLU presence has been assayed in Sjögren patients (SS) (Cuida *et al.*, 1997). Labial salivary gland biopsy specimens and saliva samples from SS expressed sCLU in both interstitial tissues and the saliva of these SS patients. In this context, sCLU could be a protector for exocrine glands of the complement-mediated injury and its function conceived as a blocker of tissue injury. sCLU is definitely a component of glomerular immune deposits in the kidney. Furthermore, increased sCLU expression also occurs in a number of renal injury states. The expression of its mRNA could be stimulated by folic acid in rats, independently of the presence of an intact complement system, because similar increases in sCLU expression were observed in C5-sufficient and C5-deficient mice (Correa-Rotter *et al.*, 1992).

What remains to be elucidated is the impact of sCLU when MAC is inhibited by streptococcal protein, such as SIC (streptococcal inhibitor of complement-mediated lysis) which is a virulence determinant of Streptococci (Akesson *et al.*, 1996).

Evaluation of endometrial tissue-specific complement activation in women with endometriosis showed that terminal C regulatory proteins (sCLU and vitronectin) were colocalized with C protein of the complement on elastic fibers (D'Cruz and Wild, 1992). On the contrary, no specific staining for C protein was detected on the glandular epithelium of uterine and ectopic endometrial tissues, suggesting a lack of specific deposition of C in the glandular epithelial cells of endometria of women with endometriosis. In endometriosis, sCLU could be a missing regulatory factor. In endometrial tissue, phytoestrogens induce complement C3 upregulation, and downregulation of CLU mRNA expression. These molecules are known for the fact that they do not induce any cell growth modification and for being a weak estrogen receptor agonist in rats. Whether it is linked to sCLU, even in part, is not yet known (Diel *et al.*, 2001).

sCLU has already been described in other nonimmune processes where chronic inflammation takes place, such as ulcerative process (Balslev *et al.*, 1999). In ischemic and/or venous hypertension leg ulcers, the deposition of complement, plasma complement regulators, and expression of membrane regulators was detected, among them vitronectin and sCLU. Staining for terminal complex complement was always associated to staining for CLU and vitronectin. However, it could not counterbalance the loss of CD59 that may enhance deposition of terminal complex of the complement.

Finally, sCLU is produced in liver stress (e.g., ethanol model in mice) with factor B, and the C1qA-chain. Despite the downregulation of MAC protein in this context, the induction of early classical and alternative pathway components together with the removal of terminal pathway components and soluble regulators have been considered as playing a role in alcohol-induced liver injuries (Bykov *et al.*, 2007).

Infection and inflammation induce alterations in hepatic synthesis and plasma concentrations of the acute-phase proteins. sCLU is secreted after endotoxin (LPS), tumor necrosis factor (TNF), and interleukin IL-1 stimulation in liver and in serum (Hardardottir *et al.*, 1994). Consequently, in some circumstances, these results demonstrated that sCLU could be regarded as protein of the acute phase of inflammation.

A strong expression of sCLU was observed on hepatic posttransplantation. It has been suggested that the strongest levels of sCLU during the reaction phase in tolerogenic induction following liver transplantation may be involved in tolerance induction (Chiang *et al.*, 2000).

B. Protective Effect of CLU Independently of Complement Activation

A protective effect of sCLU has been described independently of complement activation in kidney and heart. In renal proximal tubule cells, sCLU provided complement-independent protection against gentamicin-induced cytotoxicity but offered only marginal protection against ATP depletion-induced injury (Girton *et al.*, 2002). Moreover a protection of cardiomyocytes by sCLU independently of complement has been assayed, even if ATP was produced. The mechanism of the complement-independent cell protective effect of sCLU might differ between renal proximal tubule cells and cardiomyocytes.

C. CLU Interaction with Ig

sCLU bounds to all human IgG, IgA, and IgM isotypes with particular affinities with IgG3, IgG4, IgM, IgG1, IgG2, and lastly to IgA. Aggregated IgG in solution was a potent inhibitor of the binding of sCLU to immobilized IgG. CLU bound to both the Fab and Fc fragments of human IgG. The sCLU binding site(s) on the Fc did not overlap with those for protein A and C1q (Wilson and Easterbrook-Smith, 1992). It is also known that sCLU prepared from human serum by monoclonal antibody affinity chromatography has the ability to increase the rates of formation of insoluble immune complexes, together with C1q (Roeth and Easterbrook-Smith, 1996).

D. CLU and Immune Cells

1. DENDRITIC CELLS: AT THE CROSSROAD OF INNATE AND ADAPTATIVE IMMUNITY

During our study on CLU expression, we found that dendritic cells (DC) expressed high level of CLU and that CLU expression increased by more than 30-fold during DC maturation (unpublished personal data). Although the exact role of CLU in DC physiology is presently unknown, the main role of these cells in both arms of immunity (innate and adaptative) together with the multiple role of CLU in several processes in which DC are involved, argue for a need for better understanding of the interplay between CLU and DC.

DC are the most efficient antigen presenting cells (APC). They play a unique role in initiating immunity through the activation of naïve T cells and support local immune responses. Functions of DC are influenced by the activation status and surrounding environments. DC actively participate in the induction of both immunity and tolerance. From the plethora of receptors on DC, TLR, Fc gamma receptors Fcg-symbol gamma-R, and lectins are probably the most intensively investigated in autoimmune diseases such as RA (Bax et al., 2007; Herrada et al., 2007; Laborde et al., 2007; van Vliet et al., 2007). Upon encountering danger signals DC commence a complex process generally termed as "maturation," which involves numerous pathways and greatly enhances their capacity for antigen processing, presentation to naïve T cells and eventually, priming of T cell responses (Guermonprez et al., 2002). $Fc\gamma R$ are expressed on the cell surface of various hematopoietic cell types. They recognize IgG and IgG-containing IC and as a result constitute the link between humoral and cell-mediated immunity (Nimmerjahn and Ravetch, 2006). In man, the $Fc\gamma R$ system comprises two opposing families, the activating $Fc\gamma Rs$ I, IIa, and III and the inhibitory $Fc\gamma RIIb$, the balance of which determines the outcome of IC-mediated inflammation. Both activating and inhibiting $Fc\gamma R$ are expressed on immature DC and complete maturation of DC results in a downregulation of $Fc\gamma R$ expression. Evidence indicates that the balance between activating and inhibitory $Fc\gamma R$ determines DC behavior (Boruchov et al., 2005; Dhodapkar et al., 2007; Radstake et al., 2004), and the balance between activating and inhibitory $Fc\gamma R$ has been shown to be crucial in the susceptibility to and phenotype of various inflammatory conditions. DC also express a wide variety of C-type lectins (CLR), molecules that bind the carbohydrate moiety of glycoproteins

(Figdor *et al.*, 2002). Therefore, DC are often referred to as the "generals" of the immune system controlling several immunological checkpoints both centrally as well as peripherally. In this light, DC control many T cell responses (Gottenberg and Chiocchia, 2007; Steinman, 2007; Steinman and Banchereau, 2007). This variety in immune deviation perfectly illustrates the necessity of the DC-driven immune response for survival of the host. However, a failure to control or counteract these pathways would lead to the opposite. Namely, an uncontrolled Th-1 response leads to sepsis, shock, and possibly a breakthrough of tolerance initiating autoimmunity or cancer. In turn, an exaggerated Th-2 response leads to asthma and fibrosis, whereas a dominant IL-17 production ignites an autoimmune reaction. Interestingly, human Th-17 cells exhibited lower expression of CLU, and higher Bcl-2 expression and reduced apoptosis in the presence of TGF-β, in comparison with Th-1 cells (Santarlasci *et al.*, 2009). Moreover, human Th-17 clones and circulating Th-17 cells showed lower susceptibility to the antiproliferative effect of TGF-β than Th-1 and Th-2 clones or circulating Th-1-oriented T cells, respectively. Since most of the autoimmune diseases and cancers appear to have a Th-1 and/or Th-17 component, altogether these results reinforce the potential implication of CLU in chronic inflammation, autoimmunity, and cancer. The differentiation of DC, Th-1, and Th-17 are regulated by inflammatory signals including cytokines, PAMPs, and DAMPs that are mostly regulated through the NF-κB pathway.

E. CLU and NF-κB

Recently, it became evident that the different functions of CLU might depend on its final maturation and localization (for more details, see chapter "Clusterin (CLU): From one Gene and two transcripts to many proteins"). The predominant form is a secreted heterodimeric protein of 80 kDa (sCLU) (Murphy *et al.*, 1988) produced by translation of the full-length single mRNA encoded by a single gene located on chromosome 8 (Slawin *et al.*, 1990). sCLU is derived from a pre-sCLU protein of 60 kDa targeted to the endoplasmic reticulum and glycosylated. A nuclear form of CLU (nCLU) was recently reported to be the resulting product of an alternative splicing of exon 2 (Leskov *et al.*, 2003): please see also chapter "Nuclear CLU (nCLU) and the fate of the cell" to this regard. Finally, it recently became evident that numerous reported functions of CLU were related to intracellular forms of the protein, in particular NF-κB signaling and apoptosis. Indeed, Santilli *et al.* (2003) reported that the intracellular level of CLU could be essential for regulation of NF-κB activity via its effect on the modulation of IκB expression potentially through interaction with relevant ubiquitin ligase.

It is well known that NF-κB-induced gene expression contributes significantly to the pathogenesis of inflammatory diseases such as arthritis (Barnes and Karin, 1997; Bondeson et al., 1999; Makarov, 2001; Miagkov et al., 1998; Tak and Firestein, 2001; Tak et al., 2001), and IKK2 has been showed to be a key convergence pathway for cytokine-induced NF-κB activation in synoviocytes (Aupperle et al., 2001a,b). Nevertheless, the molecular mechanism of this activation is still largely unknown. To address this question, we generated a large number of CLU molecular constructs allowing the expression of the α and/or β chain of CLU as well as different forms or specific regions of CLU. These new tools turned out to be extremely useful to set up a functional assay to know whether CLU expression blocks NF-κB activation and nucleus translocation, and to characterize the domains involved in such an effect.

We demonstrated that CLU, like β βTrCP, the main E3 ubiquitin ligase involved in IκB degradation, interacts with phosphorylated IκBα (ppIκBα) (Devauchelle et al., 2006). We demonstrated this interaction in either HeLa or fibroblast-like synovial cells stimulated with TNF-α. IκBαpp was detectable in the anti-CLU immunoprecipitate using antibodies that recognize specifically Ser32-Ser36 ppIκBα. No interaction was detected between non-phosphorylated IκBα and CLU. These results led us to propose that the demonstrated CLU-induced IκBα stabilization (Santilli et al., 2003) could be explained by inhibiting βTrCP binding to ppIκBα. Conversely, lack of CLU expression in RA would result in enhanced IκBα degradation through βTrCP binding. As a consequence, the defect in intracellular CLU could be one key mechanism responsible for the greater activation status of NF-κB in RA synovium compared with noninflammatory joint samples.

It is presently unclear whether the interaction between CLU and phosphorylated IκBα is direct or indirect although we obtained recent evidence arguing for a direct interaction (G. Chiocchia, unpublished observation). Furthermore, it is of interest to determine whether CLU behaves as a dominant-negative form of βTrCP and whether other βTrCP substrates can bind to CLU and be protected from proteasome degradation by CLU expression.

F. Cellular Regulation Through Apoptosis

Apoptosis, defined as a programmed cell death, is a key mechanism of homeostasis. It is a controlled cell death which follows specific biochemical pathways with specific morphological changes. It allows maintaining of the balance between cell growth and cell death in a controlled pathway softer than necrosis. It is implicated in stress, such as chemicals or radiation, but also in both the development and adult homeostasis of the immune system. An apoptotic cell degrades its DNA, fragments its nucleus, shrinks its cytoplasm, and produces apoptotic bodies containing still intact structures (e.g., lysosomes, mitochondria). Apoptotic bodies are rapidly subject to phagocytosis by phagocytic cells among them DC, without the release of deadly contents (e.g., oxidative molecules, enzymes). This clever system prevents local damage produced by the release of enzyme in the local environment and stops the occurrence or development of inflammations. Furthermore, through their role as APC, DC are not only phagocytic cells but, also play a major role in antigenicity and in both central tolerogenic mechanisms and peripheral tolerogenic mechanisms. Peripheral tolerance is of particular interest in RA, as we shall explore at a later point in this publication. If apoptosis is deficient this can result in lymphoproliferative disorders or in autoimmunity occurring in the case of lack of negative selection during T cell ontogeny in thymus which favors autoimmune T cell emergence.

Numerous studies have been focused on the role of CLU in regulation of the life and death fate of cancer cells (Moretti et al., 2007; Rizzi et al., 2009; Scaltriti et al., 2004a,b; Trougakos et al., 2004). sCLU inhibits apoptosis, in human cancer cells notably, through interfering with Bax activation in mitochondria (Zhang et al., 2005). In contrast to other inhibitors of Bax, sCLU interacts with conformation-altered Bax, in the context of clinical response to antitumoral drugs. This interaction hinders Bax oligomerization, which leads to the release of cytochrome c from mitochondria and caspase activation. Moreover, sCLU inhibits oncogenic c-Myc-mediated apoptosis and promotes c-Myc-mediated transformation in vitro. Thus interfering with Bax proapoptotic activities could paradoxically favor tumor progression and inflammation. Trougakos et al. (2009) recently showed that sCLU binds and thereby stabilizes the Ku70–Bax protein complex serving as a cytosol retention factor for Bax suggesting that elevated sCLU levels may enhance tumorigenesis by interfering with Bax proapoptotic activities. On the other hand, several reports demonstrated further that the nCLU is a proapoptotic protein (Kaisman-Elbaz et al., 2009; Markopoulou et al., 2009; Rizzi et al., 2009). This point is developed in more detail in chapters "Nuclear CLU (nCLU) and the fate of the cell," in this volume and "Regulation of CLU gene expression by oncogenes, epigenetic regulation, and the role of CLU in tumorigenesis," Vol. 105.

G. From Acute to Chronic Inflammation: A Place for Oxidative Burst, Nitric Oxide

One of the key mechanisms of inflammation, especially in fighting against microbes, is respiratory burst. Oxidation is characterized by the acceptance of an electron. This induces alteration of the natural function of the protein that has accepted the electron. This explains that microbes could die when their proteins are reduced, and why oxidoreduction can lead to tissue damage. As chapter "CLU as a novel biosensor of reactive oxygen species" is directly devoted to respiratory burst and CLU, we will only present a study where DNA damage has been assayed, that could represent one mechanism of autoimmune diseases.

Bcl-2 facilitates recovery from DNA damage following oxidative stress as assayed with CLU DNA repair (Deng *et al.*, 1999). Within minutes after exposure of cells to low concentrations of hydrogen peroxide and peroxynitrite, significant mitochondrial and nuclear DNA damage is evident, with mitochondrial DNA damages greater than nuclear DNA damages. Expression of bcl-2 in PC12 cells inhibited nitric oxide donor (sodium nitroprusside)- and peroxynitrite-induced cell death. Despite the fact that oxidative insults caused both genomic and mitochondrial DNA damage in cells expressing bcl-2, recovery from DNA damage was accelerated in these cells.

V. THE PLACE AND ROLE OF CLU IN AUTOIMMUNE DISEASES (TABLE I AND FIG. 1)

A. Rheumatoid Arthritis

Rheumatoid arthritis (RA) is a chronic autoimmune and inflammatory disease that devastate articulation components among them synovial membrane, cartilage, and bone. RA is the most common chronic inflammatory joint disease, with a prevalence of 0.5% of the general population. This is a

Table I   CLU Hypothetical Mechanisms in Chronic Disease

Diseases	CLU expression	Hypothetical mechanism	References
Lupus	Low	Low complement inhibition	(Moll *et al.*, 1998), (Newkirk *et al.*, 1999)
Rheumatoid arthritis	Low	Apoptosis Regulator of NF-κB	(Connor *et al.*, 2001), (Devauchelle *et al.*, 2006)
RA model	Normal	Autoantigen?	(Kuhn *et al.*, 2006)
Juvenile dermatomyositis	High	Unknown	(Chen *et al.*, 2008)
Diabetes	High	Marker of disease Protector or pathogenic regarding the subcellular localization	(Calvo *et al.*, 1998b), (Savkovic *et al.*, 2007)

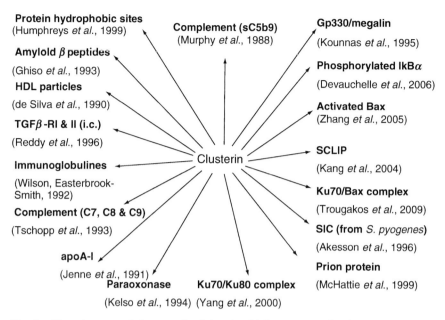

Fig. 1 Clusterin a central rheostat of activity of multiple proteins and pathways involved in autoimmune reaction. Molecular interaction of CLU linked to chronic inflammation and autoimmunity.

frequent chronic disease which reduces life span of 10 years. Environmental factors are usually discussed as well as genetic susceptibility (Jaen *et al.*, 2009), mainly linked to TNF-α and antigen presentation (Lutzky *et al.*, 2007; Olsen and Stein, 2004). Whether T cells play the central role in the development of RA is not definitely admitted. Some authors hypothesize that the main culprits could be the activation of non-T cells (most notably macrophage-like and fibroblast-like synovial cells, FLS) and proinflammatory molecules (Chiocchia *et al.*, 1993, 1994; Firestein and Zvaifler, 1990). In this hypothesis, the antigen-specific response is of lesser importance. It highlights the uncertainties surrounding the role for T cells and autoantigens (Firestein and Zvaifler, 2002). The synovial cells from the lining and sublining layers are activated and hyperproliferative (Fujinami *et al.*, 1997). Among the identified genes mainly modulated in RA synovitis, CLU is of interest because of its multiple functions related to apoptosis, inflammation, proliferation, and differentiation (Devauchelle and Chiocchia, 2004; Devauchelle *et al.*, 2004, 2006). Indeed in synovial tissue, the protein is predominantly expressed by synoviocytes and is detected in synovial fluids. Both full-length and spliced isoform CLU mRNA levels of expression seem to be lower in RA tissues compared with osteoarthritis (OA) and healthy synovium. On the other hand, the intracellular form of CLU is merely detectable. The overexpression of CLU (both sCLU and nCLU forms) in RA-FLS promoted apoptosis within 24 h and raised the possibility that it could participate in a mechanism of homeostasis in RA.

Cytokines have a heavy impact on the development and on the clinical course of RA: blocking TNF-α, IL-6 or IL-1 demonstrated that these cytokines play a major role in the disease in which they have different implications. Indeed, inhibiting the actions of these cytokines by means of antibodies or natural biological inhibitors can even lead to remission. We observed that CLU knockdown with small-interfering RNA promoted IL-6 and IL-8 production by cytokines directly involved in inflammation (Devauchelle et al., 2006; Falgarone et al., submitted for publication). CLU interacted with phosphorylated IκBα and could also be responsible for an enhancement of NF-κB activation and survival of the synoviocytes. On the other hand, CLU knockdown does not modify the expression of other cytokines known to be highly expressed in RA-FLS such as IL-32 and MCP-1 (Cagnard et al., 2005).

Because the angiogenesis is a key mechanism of RA synovitis, CLU assays in the context of angiogenesis are of particular interest. The antiangiogenic properties of CLU antisense nucleotide (ASN) using a capillary cell (HUVEC) viability assay have been studied (Jackson et al., 2005). CLU ASN's and bcl-2, bcl-xl and survivin ASN's were found to inhibit HUVEC growth. Combinations of ASN's with drugs known to inhibit HUVEC growth demonstrated a minor additive but not synergistic inhibitory effect on HUVEC proliferation. CLU ASN's were found to strongly inhibit angiogenesis and induce high levels of apoptosis in HUVECs. It is of interest to notice that a strong antiangiogenic action of CLU ASN's, is not necessarily related to any chemosensitization effect of this agent (Lourda et al., 2007) (Please see also chapter "CLU and chemoresistance," Vol. 105 for more details on CLU and chemoresistance).

However, the role of CLU in RA linked to complement inhibition is not excluded. Indeed, we used the K/BN model of arthritis where arthritis develops rapidly after the injection of serum of KRNxNOD mice that secrete autoantibodies directed against GPI (glucose-6-phosphate isomerase). In this model, arthritis is the consequence of antibody deposits on tissue that directly activate the complement and cause arthritis inflammation (Matsumoto et al., 2002). We injected antibodies and induced arthritis in CLU KO mice (McLaughlin et al., 2000) and syngenic CLU wild-type mice. We observed that arthritis developed faster in CLU KO mice, 1 day before wild-type model, and lasted longer in CLU KO mice compared with CLU wild-type mice (Fig. 2), even if we did not observe any significant statistical difference in arthritis score. Interestingly, the CLU-deficient animals were also showed to be more susceptible to autoimmune myocarditis (McLaughlin et al., 2000).

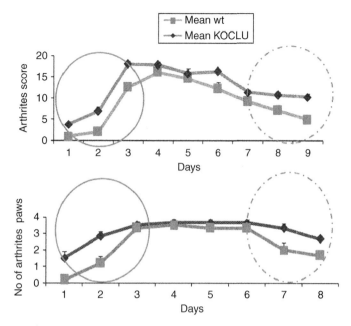

Fig. 2 Absence of CLU results in an accelerated and more lasting arthritis. C57Bl/6 CLU wild-type or CLU KO mice received either 2×100 μl of K/BxN arthritic mice. First signs of arthritis appeared between day 2 and 3 for WT mice compared to 1 and 2 in KO animals. The severity of the disease was the same in both groups but the symptoms persisted longer.

CLU was measured in synovial fluid simultaneously than C5a and vitronectin. In RA, the concentration of C5a is increased threefold in plasma (21.9 vs. 7.2 μg/l) and fivefold in SF (7.8 vs. 1.7 μg/l) compared to OA. The SF/plasma ratios for C5a, vitronectin, and CLU were 0.35, 0.36, and 0.23, respectively, not significantly different between the two diseases. This brings out the conclusion that the high level of complement terminal pathway activation in RA-SF combined with low local levels of lysis inhibitors might allow lytic or sublytic attacks on local cells, resulting in inflammation and cell damage (Hogasen *et al.*, 1995).

Studies involving autoantibodies have led to the identification of many potential autoantigens in RA (Corrigall and Panayi, 2002) including antigens expressed in the joint (e.g., type 2 collagen, chondrocyte glycoprotein gp39, and proteoglycans), antigens expressed only at nonarticular sites, and ubiquitous antigens (e.g., heat-shock proteins and glucose-6-phosphate isomerase). Recently, the role for posttranslational modification of proteins has been enlightened by many studies (Doyle and Mamula, 2001; Rudd *et al.*, 2001) and by the clinical implication of antibody directed against citrullinated peptides. These considerations are based on the formation of new epitopes called neoepitopes which become the target of autoantibodies. Citrullination of arginine residues in proteins has been incriminated (Masson-Bessiere *et al.*, 2001; van Gaalen *et al.*, 2004) in the loss of self-tolerance in RA; antibodies directed against citrullinated peptides are definitely crucial antibodies for both diagnosis and prognosis in RA. Interestingly, it was recently reported in collagen-induced arthritis, a murine model of RA (Boissier *et al.*, 1989; Courtenay *et al.*, 1980), that anti-CLU citrullinated antibodies could react with CLU (Kuhn *et al.*, 2006).

T cell hybridoma clones specific for glycosylated epitopes were derived from HLA-DR4 transgenic mice expressing HLA-DRB*0401, a class II antigen associated with increased severity (Backlund *et al.*, 2002) and susceptibility of human RA. Among these mechanisms, cell–cell interaction plays a special part because it seems to occur in synovial membrane where the formation of germinal centers has been described (Weyand *et al.*, 2000).

B. Cellular Dysregulation in Synovitis

Synovitis is characterized by an imbalance in favor of proinflammatory molecules followed by deleterious activation of T cells, which are predominantly Th-1 and Th-17 cells in this environment (Boissier *et al.*, 2008; Feldmann *et al.*, 1996). T cells contribute to about 30% of the cell infiltrate and are found in a perivascular location, suggesting migration from the peripheral blood to the synovial membrane. T cell recruitment to the synovial membrane involves interactions with the endothelial cells lining the synovial high endothelial venules (Page *et al.*, 2002). T cells contribute to potentiate the autoimmune response and to activate B cells (thereby promoting autoantibody production) in cooperation with macrophages, dendritic cells, synovial cells, and fibroblasts. In addition, T cells interact with osteoclasts and ultimately promote the development of an autoimmune response, the production of autoantibodies, angiogenesis, and the destruction of bone and cartilage.

Incontrovertible proof that T cells are involved in RA was obtained in studies of mutant mice lacking mature T cells because these animals are refractory to collagen-induced arthritis.

DC present autoantigens and maintain peripheral self-tolerance. Their role in self-tolerance depends closely on their state of maturation or activation (Steinman *et al.*, 2003). A T cell-dependent disease necessarily involves cell–cell interactions and therefore APC; and the autoimmune B cell response is amplified by the T cell response. We have learned from animal models, as well as from recent clinical data available on this disease, that T cells and/or regulatory cells are key actors, even if their own impact has still to be elucidated. T cells that express a memory phenotype CD4+ CD45RO+ were identified in the rheumatoid synovial membrane (Kohem *et al.*, 1996).

A distortion in the T cell repertoire has been found in patients with RA and may stem from an abnormality in thymic output that may allow the release of autoreactive T cells exhibiting strong affinity for autoantigens (Goronzy and Weyand, 2001).

The mechanisms by which T cells undergo activation are controlled by effector cells.

Alterations in CD4+ CD25bright Treg cells were found in blood and joint specimens from patients with RA (Cao *et al.*, 2003). Similarly, deficient function of peripheral blood Treg cells from patients with RA was demonstrated (Ehrenstein *et al.*, 2004). The Treg cells suppressed the proliferation of effector T cells but failed to inhibit their ability to secrete proinflammatory cytokines (Ehrenstein *et al.*, 2004).

Among cytokines that are overexpressed in RA, IL-6 is one of the major pathologic factors.

CLU has been recently demonstrated to be under the regulation of IL-6, in the cancer context (Pucci *et al.*, 2009), but with strong echo for RA pathophysiology. IL-6 seems to affect prosurvival pathways in colon cancer progression, modulating the expression and the molecular interactions among the proapoptotic factor Bax, the DNA repair proteins Ku70/86 and CLU forms. Hyper IL-6 production could induce the loss of Ku86 and CLU 50–55 kDa proapoptotic form, while Bax and the 40 kDa prosurvival sCLU form appeared to be expressed. IL-6 downmodulated Bax RNA expression, and IL-6 exposure influenced Bax protein level acting on the Bax–Ku70–sCLU physical interactions in the cytoplasm. The Ku70 acetylation and phosphorylation state are then modified and this could lead to the inhibition of Bax proapoptotic activity. In addition, IL-6 treatment induced a significant downregulation of Ku86 and a strong increase of sCLU. In contrast, somatostatin treatment of Caco-2 cells was able to restore apoptosis, demonstrating that Ku70–Bax–CLU interactions could be dynamically modulated in tumor cells. Hence, IL-6 could promote cell survival and apoptosis escape. In this line, it is also of importance to note that enhanced Bcl-2 expression and NF-κB nuclear translocation of synovial cells are induced by inflammatory cytokines and/or growth factors, and that synovial cells become resistant toward apoptosis triggered by various stimuli. Clearly, deciphering the exact role of CLU in all these processes is the next challenge.

C. Cartilage

Following this work, Connor *et al.* (Connor *et al.*, 2001; Kumar *et al.*, 2001) studied *clu* mRNA expression in adult human normal and OA cartilage. They demonstrated the abundance of *clu* mRNA being similar to that of protein known to be frequently expressed in cartilage. The expression of *clu* mRNA was upregulated in early OA versus normal cartilage when analyzed by microarray analysis, but *in situ* hybridization showed moderate levels of CLU in chondrocytes of normal cartilage. In histological studies, sCLU appeared to be produced in superficial zone of cartilage (Khan *et al.*, 2001). Its expression is upregulated during maturation but is confined to the articular surface. The induction of sCLU seemed to be associated with a variety of disease states where it appears to provide a cytoprotective effect (Connor *et al.*, 2001). The increased expression of *clu* mRNA in the early stages of OA was considered to be an attempt by the chondrocytes to protect and repair the tissue. In contrast, sCLU decrease in advanced OA cartilage could be the sign of the degenerative stage of the disease. Since, it has been hypothesized that *clu* has a role in cell differentiation these results call to answer to two very important questions: First, is CLU involved in chondrocyte differentiation; and if yes, which form of CLU; and second, does CLU has a pathogenic role in OA? Clearly, answering these questions will be a major challenge.

In cartilage cultures, CLU appears among the proteins lost by explants in culture medium containing IL-1β and TNF-α (Stevens *et al.*, 2008). IL-1α and TNF-α cause increased release of chitinase 3-like protein 1 (CHI3L1), CHI3L2, complement factor B, matrix metalloproteinase 3, ECM-1, haptoglobin, serum amyloid A3, and CLU. In contrast, injurious compression causes the release of intracellular proteins, including Grp58, Grp78, α-4-actinin, pyruvate kinase, and vimentin.

Finally CLU has been purified in complex with a metalloproteinase, named MP6-MMP or MMP25 (Matsuda *et al.*, 2003). CLU inhibits the activity of this enzyme but not that of MMP-2 or MT1-MMP. MMP25 could be involved in cancer metastatis, but also in OA (Hopwood *et al.*, 2007). The role of this interaction in RA is presently unknown.

D. CLU at the Crossroad of Autoimmune Reactions in the Joint

Helper T cells perform critical functions in the immune system through the production of distinct cytokine profiles. In addition to Th-1 and Th-2 cells, a third subset of polarized effectors T cells characterized by the production of IL-17 and other cytokines (and now called Th-17 cells) is associated with the pathogenesis of several autoimmune conditions (Furuzawa-Carballeda *et al.*, 2007). Furthermore, naïve T cells can be converted into regulatory T (Treg) cells that prevent autoimmunity in presence of the cytokine transforming growth factor-β (TGF-β). However, in the presence of IL-6, TGF-β has also been found to promote the differentiation of naïve T lymphocytes into proinflammatory IL-17 cytokine-producing Th-17 cells, which promote autoimmunity and inflammation.

Although the factors that promote murine Th-17 differentiation have been intensively investigated, there has been much less information on the regulation of the IL-17 cytokine production in human T cells than in mice. IL-17 is readily produced by human memory T cells (Chen *et al.*, 2007; Wilson *et al.*, 2007). Interestingly, human Th-17 cells exhibited lower expression of CLU, and higher Bcl-2 expression and reduced apoptosis in the presence of TGF-β, in comparison with Th-1 cells (Santarlasci *et al.*, 2009). These T cells exhibit distinct patterns of chemokine receptor expression such as CCR6, CCR4, and CXCR3 which are important for localization to tissues (Acosta-Rodriguez *et al.*, 2007). Furthermore, chemokine receptor expression seems to discriminate specific pattern of cytokine production by Th-17 cells namely: IFN-γ-IL-17 dual producers or IL-17-specific producers, and we know that this type of dual producer T cells is present in RA joint. What is the level of expression of CLU and its role in these cells has to be determined.

It was recently showed that the differentiation of human naïve CD4+ T cells into Th-17 cells is promoted by IL-1β and IL-6. Furthermore, it has been shown that IL-1, IL-6, and IL-23 are important in driving human Th-17 differentiation. However, TGF-β1 which is important for the differentiation of murine Th-17 cells and inducible regulatory T cells (iTregs) is reportedly not required and even inhibits human Th-17 differentiation. In addition, human Th-17 cells also produce other proinflammatory cytokines. Chung et al. reported the interplay between IL-6 and IL-1 through induction of IL-1R by IL-6 for the differentiation of Th-17 cells (Chung *et al.*, 2009). The multiple roles of IL-1, IL-6, and IL-23 in the Th-17 axis are important because they offer opportunities for therapeutic interventions. In particular IL-6 appears to be a key cytokine in this TH pathway differentiation and is also common to both mice and human (Acosta-Rodriguez *et al.*, 2007). Finally, Liu's group reported that a subpopulation of human Treg based on the CD4+ CD25high phenotype was able to produce IL-17 (Voo *et al.*, 2009). Quantification of such population in diseased state and progression of this T cell number following treatment has never been studied.

We already discussed the potential role of CLU modulation of expression on TLR signaling and IL-6 expression and production. Because IL-1R, like TLR, operates through MyD88 to activate NF-κB and IL-6 acts through NF-κB either and because synoviocytes produced higher amount of IL-6 and IL-8 when CLU is downregulated, it is tempting to speculate on the role of CLU in the immune dysregulation occurring in RA synovitis. The unanswered question is: what is the mechanism inducing downregulation of CLU expression in RA synovium?

E. Systemic Lupus Erythematosus

SLE is a complex autoimmune disease which etiopathogeny involves high titers of autoantibodies directed against various antigens including nuclear as well as cytoplasmic antigens. This disease is characterized by chronic inflammation mainly mediated by antibodies and immune complexes of various organs leading to glomerulonephritis, dermatitis, arthritis, serosities, and vasculitis. It is a rare disease estimated at 0.05% of the general population, with a clear female sex ratio (9:1). Both environmental and genetic factors contribute to the disease physiopathogenesis, among them it should be noted that Afro-American or Hispanic people are more likely to develop the disease. This disease is a polygenic disease, but C1 and C4 deficiency leads to the development of SLE.

SLE is also characterized by multisystem microvascular inflammation with the generation of autoantibodies. It has been hypothesized that the redistribution of cellular antigens during apoptosis leads to a cell-surface display of plasma and nuclear antigens in the form of nucleosomes. Thus, dysregulated B and T lymphocytes begin targeting normally protected (hidden) intracellular antigens. In SLE, both T and B cells are hyperreactive and many immunoregulatory frameworks are defective. It has been reported that B cells are subject to polyclonal and specific activation, driven by cytokine production (e.g., IL-6) while IL-10 can prevent mice lupus (Finck *et al.*, 1994).

The T cell disturbance is less clear in human SLE. If cellular counts are lower, with less proliferation of T cells to mitogenic molecules, some consider that they are the key cells for antibody production enhancement. Indeed many antibodies are IgG, and need T helper interplay to stimulate antibody production by B cells.

Lupus being a chronic disease, the production of pathogenic subsets of autoantibodies and immune complexes must take place along with a lack of downregulation. In SLE, both autoreactive T and B cells circulate in peripheral blood. A defective apoptosis is supported by the *lpr* and *gld* mutation (in lpr/mrl mice) responsible for the lupus-like disease in the strain of mice. The above-quoted abnormalities lead to inadequate clearance of immune complexes. They are less cleared by macrophages and deposit in tissues. One typical clinical expression is the nephritis deposits of autoantibodies which are accompanied by immune complex deposits and complement activation in the kidneys.

F. CLU Implication in SLE

CLU which is constitutively expressed in tubular epithelial cells displays an enhanced expression following tubular injuries and it has been detected in glomerular immune deposits of glomerulonephritis. In lupus-like nephritis, a significant increase of CLU mRNA abundance was demonstrated (Moll *et al.*, 1998). This upregulation was localized exclusively in tubular epithelial cells exhibiting tubulointerstitial alterations, whereas no CLU mRNA was detectable in diseased glomeruli, excluding an active synthesis of CLU by glomerular cells. A similar tubular increase of CLU mRNA abundance was observed in myeloma-like cast nephropathy induced by IgG3 monoclonal cryoglobulins and even though there were no detectable histological alterations in a model of septic shock induced by the injection of bacterial lipopolysaccharides. Indeed, the tubular epithelial cells are the only sites of CLU mRNA accumulation during the course of lupus-like nephritis. Furthermore, the tubular upregulation of CLU gene expression may reflect the cellular response to various types of tubular injuries.

CLU role in kidney aging is studied with a view to deciphering effects of diminished protection. Aging mice deficient in CLU develop a progressive glomerulopathy characterized by the deposition of immune complexes in the mesangium (Rosenberg *et al.*, 2002). Up to 75% of glomeruli in CLU-deficient mice exhibit moderate to severe mesangial lesions by 21 months of age. This being said, wild-type and hemizygous mice exhibited little or no glomerular pathology. In the CLU-deficient mice, immune complexes of immunoglobulin G (IgG), IgM, IgA, and sometime C1q, C3, and C9 are detectable as early as 4 weeks of age. The accumulation of material can be found in the mesangial matrix and age-dependent formation of intramesangial tubulofibrillary structures. Interestingly, damaged glomeruli showed no evidence of inflammation or necrosis. In young CLU-deficient animals, the development of immune complex lesions can be accelerated by unilateral nephrectomy-induced hyperfiltration. The above data reinforce the protective role of CLU against chronic glomerular kidney disease and support the hypothesis that CLU modifies immune complex metabolism and disposal.

In patients with SLE, the serum levels of sCLU correlate with disease (Newkirk *et al.*, 1999). The disease activity scores of 80 patients (mainly women) were determined using the systemic lupus activity measure-revised and significant decrease in patients with SLE was observed that correlated inversely with disease activity. Low sCLU levels were significantly associated with skin ulcers, loss of hair, proteinuria, and low platelet count, and strikingly associated with arthritis. Interestingly, the sCLU levels did not correlate with either systemic complement consumption, as measured by C3 or C4, or with prednisone use. It is overwhelmingly hypothesized that individuals lacking sufficient amounts of sCLU could have poor control of antibody-mediated inflammation at sites of apoptosis where autoantigens are exposed.

Cardiovascular complications are common in SLE and myocardial infarctions are considered to be the leading cause of increased mortality. The ADP receptor P2Y12 plays a central role in platelet activation and the P2Y12 blocker clopidogrel reduces the incidence of cardiovascular events. As CLU has been assayed by microarray study to be expressed at very high levels in platelets, it has been studied in lupus platelets (Wang *et al.*, 2004). SLE patients had significantly lower levels of CLU than controls ($P < 0.05$) while platelet aggregation was similar in both groups and despite a difference in P2Y receptor. Lowered sCLU levels could be involved in the pathogenesis of SLE due to the decrease of protective effects among them low platelet activator. Direct evidence is still missing at this stage.

G. Diabetes

At the opposite of RA and SLE, sCLU is overexpressed in pancreas during the acute phase of pancreatitis and seems to be protective during inflammation of exocrine pancreas because of its antiapoptotic and anti-inflammatory functions (Savkovic *et al.*, 2007). sCLU-overexpressing AR4-2J cells, pancreas acinar cells, showed higher viability after cell stress and accordingly reduced rates of apoptosis and lessened caspase-3 activation. On the contrary, blockage of endogenous CLU expression reduced viability and enhanced apoptosis. Presence of CLU reduced NF-κB activation and expression of the NF-κB target genes TNF-α and MOB-1 under cell stress in accordance with the inhibitor role of CLU on NF-κB. CLU-deficient mice showed a more severe course of acute experimental pancreatitis with enhanced rates of apoptosis and inflammatory cell infiltration (Savkovic *et al.*, 2007). This has been demonstrated in another model of pancreatitis (Calvo *et al.*, 1998a). The authors demonstrated that *clu* mRNA was expressed with similar intensity in edematous caerulein-induced pancreatitis and in response to various degrees of necrohemorrhagic taurocholate-induced pancreatitis, indicating a maximal gene activity in all types of pancreatitis. *clu* mRNA levels were strongly induced in pancreatic acinar AR4-2J cells in response to various apoptotic stimuli (i.e., cycloheximide, staurosporine, ceramide, and H2O2) but not with IL-1, IL-4, or IL-6 or heat shock, which do not induce apoptosis in these pancreatic cells. sCLU has a strong expression during acute pancreatitis and could be involved in the control of acinar cell apoptosis.

Diabetic nephropathy shows histological specificity for the implications of CLU. sCLU and nCLU have a different role in diabetic nephropathy (Tuncdemir and Ozturk, 2008). While sCLU is increased in the glomeruli and tubuli of untreated diabetic rats, an increase in the nCLU immunoreactivity observed in the podocytes, mesangial cells, and the injured tubule cells of the same rats. nCLU staining decreases in all treated diabetic rats, for example, by angiotensine convertase enzyme inhibitor. Furthermore, nCLU shares the same distribution with apoptotic cells in diabetic rats. This study shows that, in the context of diabetic nephropathy, sCLU is induced by renal tissue damage, while nCLU is related to apoptosis.

Besides these data, proteomic assays led Metz *et al.* (2008) to propose five markers to identify diabetic patients. Among the five identified proteins, three were upregulated (α-2-glycoprotein 1, corticosteroid-binding globulin, and lumican) whereas two (sCLU and serotransferrin) were twofold decreased in diabetic patients.

H. Juvenile Dermatomyositis

In juvenile dermatomyositis, genes involved in immune responses and vasculature remodeling are expressed at a higher level in muscle biopsies from children (Chen *et al.*, 2008). In these tissues, sCLU, HLA-DQA1, smooth muscle myosin heavy chain, plexin D1, and tenomodulin were upregulated. The chronic inflammation in patient with longer disease duration was also associated with increased DC-LAMP+ and BDCA2+ mature dendritic cells, identified by immunohistochemistry. It was noticed that symptoms lasting 2 months or more were associated with DC maturation and antiangiogenic vascular remodeling. How important is the upregulation of CLU in this phenomenon has now to be addressed.

I. Myocarditis

In a seminal study, it was demonstrated that KO ApoJ/CLU mice developed inflammatory myocarditis that shares common features with the wild-type mice injected model (McLaughlin *et al.*, 2000). Mice injected with myosin developed a T-dependant myocarditis that linked to both MHC class II expression and TNF-α production. This disease is strongly linked to the secondary antibody response. It was stroking that recovery following disease was deeply modified in ApoJ/CLU KO mice. These animals exhibited cardiac function impairment and severe myocardial scarring. These results led to the conclusion that CLU could limit progression of autoimmune myocarditis and protect the heart from postinflammatory tissue destruction. Thus, in this case, CLU could be a protein required for recovery process.

VI. CONCLUDING REMARKS

It is now evident that CLU can act directly and indirectly in numerous ways related to inflammation and immunity: (i) Regulation of complement activation; (ii) negative regulator of NF-κB; (iii) pro/antiapoptotic activity following isoforms; (iv) modulation of cell differentiation; (v) bidirectional regulation of and by and major proinflammatory cytokines such as TNF-α and IL-6.

sCLU is an inhibitor of complement activation and protects cells or tissues from MAC. sCLU is also secreted after endotoxin (LPS), TNF, and IL-1 stimulation. These properties of sCLU have a strong impact in autoimmune diseases.

It is now evident that the different functions of CLU might depend on its final maturation and localization. Thus numerous functions of CLU were related to intracellular forms of the protein, in particular NF-κB signaling and apoptosis.

One of the recent new functions described for CLU is its ability to induce IκBα stabilization. Thus, the defect in intracellular form of CLU observed in pathological conditions could be one key mechanism responsible for the greater activation status of NF-κB in several inflammatory conditions such as it occurs in RA synovium.

Because IL-1R, like TLR, operates through MyD88 to activate NF-κB and IL-6 acts through NF-κB either and because cells like synoviocytes produced higher amount of IL-6 and IL-8 when CLU is downregulated, it is tempting to speculate on the role of CLU in the immune deregulation occurring in inflammatory conditions and autoimmunity.

Recent data on showing that human Th-17 cells exhibited lower expression of CLU, and higher Bcl-2 expression and reduced apoptosis in the presence of TGF-β, in comparison with Th-1 cells have now open a new field of research to study the potential implication of CLU in chronic inflammation, autoimmunity, and cancer.

The modulation of proinflammatory cytokine production such as IL-6 and IL-8 occuring after CLU dysregulation and the role of CLU as a negative regulator of NF-κB argue for a prominent role of CLU in inflammatory conditions.

All of these results have now opened the gate for new questions:

What are the mechanisms inducing downregulation of CLU expression in RA synovium? What is the exact role of *clu* in cell differentiation (such as for Th-17 cells) and which forms of CLU are involved in this process? How important are the posttraductionnal modification of CLU for autoantibodies recognition? What is the exact role of CLU in regulation of apoptosis, drug resistance, and proinflammatory response of the targeted cells in autoimmune reaction?

Thus, it is clear that CLU has not revealed yet all its secrets in inflammation and autoimmunity and numerous questions are still waiting for answer and represent the next challenges.

REFERENCES

Acosta-Rodriguez, E. V., Rivino, L., Geginat, J., Jarrossay, D., Gattorno, M., Lanzavecchia, A., Sallusto, F., and Napolitani, G. (2007). Surface phenotype and antigenic specificity of human interleukin 17-producing T helper memory cells. *Nat. Immunol.* **8**, 639–646.

Akesson, P., Sjoholm, A. G., and Bjorck, L. (1996). Protein SIC, a novel extracellular protein of *Streptococcus pyogenes* interfering with complement function. *J. Biol. Chem.* **271**, 1081–1088.

Aronow, B. J., Lund, S. D., Brown, T. L., Harmony, J. A., and Witte, D. P. (1993). Apolipoprotein J expression at fluid-tissue interfaces: Potential role in barrier cytoprotection. *Proc. Natl. Acad. Sci. USA* **90**, 725–729.

Aupperle, K., Bennett, B., Han, Z., Boyle, D., Manning, A., and Firestein, G. (2001a). NF-kappa B regulation by I kappa B kinase-2 in rheumatoid arthritis synoviocytes. *J. Immunol.* **166**, 2705–2711.

Aupperle, K. R., Yamanishi, Y., Bennett, B. L., Mercurio, F., Boyle, D. L., and Firestein, G. S. (2001b). Expression and regulation of inducible IkappaB kinase (IKK-i) in human fibroblast-like synoviocytes. *Cell Immunol.* **214**, 54–59.

Backlund, J., Carlsen, S., Hoger, T., Holm, B., Fugger, L., Kihlberg, J., Burkhardt, H., and Holmdahl, R. (2002). Predominant selection of T cells specific for the glycosylated collagen type II epitope (263–270) in humanized transgenic mice and in rheumatoid arthritis. *Proc. Natl. Acad. Sci. USA* **99**, 9960–9965.

Balslev, E., Thomsen, H. K., Danielsen, L., Sheller, J., and Garred, P. (1999). The terminal complement complex is generated in chronic leg ulcers in the absence of protectin (CD59). *Apmis* **107**, 997–1004.

Barnes, P. J., and Karin, M. (1997). Nuclear factor-kappaB: A pivotal transcription factor in chronic inflammatory diseases. *N. Engl. J. Med.* **336**, 1066–1071.

Bax, M., Garcia-Vallejo, J. J., Jang-Lee, J., North, S. J., Gilmartin, T. J., Hernandez, G., Crocker, P. R., Leffler, H., Head, S. R., Haslam, S. M., Dell, A., and van Kooyk, Y. (2007). Dendritic cell maturation results in pronounced changes in glycan expression affecting recognition by siglecs and galectins. *J. Immunol.* **179**, 8216–8224.

Boissier, M. C., Chiocchia, G., and Fournier, C. (1989). [Clinical and experimental anticollagen autoimmunity]. *C. R. Seances Soc. Biol. Fil.* **183**, 293–306.

Boissier, M. C., Assier, E., Falgarone, G., and Bessis, N. (2008). Shifting the imbalance from Th1/Th2 to Th17/treg: The changing rheumatoid arthritis paradigm. *Joint Bone Spine* **75**, 373–375.

Bondeson, J., Foxwell, B., Brennan, F., and Feldmann, M. (1999). Defining therapeutic targets by using adenovirus: Blocking NF-kappaB inhibits both inflammatory and destructive mechanisms in rheumatoid synovium but spares anti-inflammatory mediators. *Proc. Natl. Acad. Sci. USA* **96**, 5668–5673.

Boruchov, A. M., Heller, G., Veri, M. C., Bonvini, E., Ravetch, J. V., and Young, J. W. (2005). Activating and inhibitory IgG Fc receptors on human DCs mediate opposing functions. *J. Clin. Invest.* **115**, 2914–2923.

Bykov, I., Junnikkala, S., Pekna, M., Lindros, K. O., and Meri, S. (2007). Effect of chronic ethanol consumption on the expression of complement components and acute-phase proteins in liver. *Clin. Immunol.* **124**, 213–220.

Cagnard, N., Letourneur, F., Essabbani, A., Devauchelle, V., Mistou, S., Rapinat, A., Decraene, C., Fournier, C., and Chiocchia, G. (2005). Interleukin-32, CCL2, PF4F1 and GFD10 are the only cytokine/chemokine genes differentially expressed by *in vitro* cultured rheumatoid and osteoarthritis fibroblast-like synoviocytes. *Eur. Cytokine Netw.* **16**, 289–292.

Calvo, E. L., Mallo, G. V., Fiedler, F., Malka, D., Vaccaro, M. I., Keim, V., Morisset, J., Dagorn, J. C., and Iovanna, J. L. (1998a). Clusterin overexpression in rat pancreas during the acute phase of pancreatitis and pancreatic development. *Eur. J. Biochem.* **254,** 282–289.

Calvo, R. M., Forcen, R., Obregon, M. J., Escobar del Rey, F., Morreale de Escobar, G., and Regadera, J. (1998b). Immunohistochemical and morphometric studies of the fetal pancreas in diabetic pregnant rats. Effects of insulin administration. *Anat. Rec.* **251,** 173–180.

Cao, D., Malmstrom, V., Baecher-Allan, C., Hafler, D., Klareskog, L., and Trollmo, C. (2003). Isolation and functional characterization of regulatory CD25brightCD4+ T cells from the target organ of patients with rheumatoid arthritis. *Eur. J. Immunol.* **33,** 215–223.

Chauhan, A. K., and Moore, T. L. (2006). Presence of plasma complement regulatory proteins clusterin (Apo J) and vitronectin (S40) on circulating immune complexes (CIC). *Clin. Exp. Immunol.* **145,** 398–406.

Chen, Z., Tato, C. M., Muul, L., Laurence, A., and O'Shea, J. J. (2007). Distinct regulation of interleukin-17 in human T helper lymphocytes. *Arthritis Rheum.* **56,** 2936–2946.

Chen, Y. W., Shi, R., Geraci, N., Shrestha, S., Gordish-Dressman, H., and Pachman, L. M. (2008). Duration of chronic inflammation alters gene expression in muscle from untreated girls with juvenile dermatomyositis. *BMC Immunol.* **9,** 43.

Chiang, K. C., Goto, S., Chen, C. L., Lin, C. L., Lin, Y. C., Pan, T. L., Lord, R., Lai, C. Y., Tseng, H. P., Hsu, L. W., Lee, T. H., Yokoyama, H., *et al.* (2000). Clusterin may be involved in rat liver allograft tolerance. *Transpl. Immunol.* **8,** 95–99.

Chiocchia, G., Manoury, B., Boissier, M. C., and Fournier, C. (1993). T cell-targeted immunotherapy in murine collagen-induced arthritis. *Clin. Exp. Rheumatol.* **11**(Suppl. 9), S15–S17.

Chiocchia, G., Manoury-Schwartz, B., Boissier, M. C., Gahery, H., Marche, P. N., and Fournier, C. (1994). T cell regulation of collagen-induced arthritis in mice. III. Is T cell vaccination a valuable therapy? *Eur. J. Immunol.* **24,** 2775–2783.

Choi, N. H., Mazda, T., and Tomita, M. (1989). A serum protein SP40,40 modulates the formation of membrane attack complex of complement on erythrocytes. *Mol. Immunol.* **26,** 835–840.

Chung, Y., Chang, S. H., Martinez, G. J., Yang, X. O., Nurieva, R., Kang, H. S., Ma, L., Watowich, S. S., Jetten, A. M., Tian, Q., and Dong, C. (2009). Critical regulation of early Th17 cell differentiation by interleukin-1 signaling. *Immunity* **30,** 576–587.

Connor, J. R., Kumar, S., Sathe, G., Mooney, J., O'Brien, S. P., Mui, P., Murdock, P. R., Gowen, M., and Lark, M. W. (2001). Clusterin expression in adult human normal and osteoarthritic articular cartilage. *Osteoarthr. Cartil.* **9,** 727–737.

Correa-Rotter, R., Hostetter, T. H., Nath, K. A., Manivel, J. C., and Rosenberg, M. E. (1992). Interaction of complement and clusterin in renal injury. *J. Am. Soc. Nephrol.* **3,** 1172–1179.

Corrigall, V. M., and Panayi, G. S. (2002). Autoantigens and immune pathways in rheumatoid arthritis. *Crit. Rev. Immunol.* **22,** 281–293.

Courtenay, J. S., Dallman, M. J., Dayan, A. D., Martin, A., and Mosedale, B. (1980). Immunisation against heterologous type II collagen induces arthritis in mice. *Nature* **283,** 666–668.

Cuida, M., Legler, D. W., Eidsheim, M., and Jonsson, R. (1997). Complement regulatory proteins in the salivary glands and saliva of Sjogren's syndrome patients and healthy subjects. *Clin. Exp. Rheumatol.* **15,** 615–623.

D'Cruz, O. J., and Wild, R. A. (1992). Evaluation of endometrial tissue specific complement activation in women with endometriosis. *Fertil. Steril.* **57,** 787–795.

de Silva, H. V., Stuart, W. D., Duvic, C. R., Wetterau, J. R., Ray, M. J., Ferguson, D. G., Albers, H. W., Smith, W. R., and Harmony, J. A. (1990). A 70-kDa apolipoprotein designated ApoJ is a marker for subclasses of human plasma high density lipoproteins. *J. Biol. Chem.* **265,** 13240–13247.

Deng, G., Su, J. H., Ivins, K. J., Van Houten, B., and Cotman, C. W. (1999). Bcl-2 facilitates recovery from DNA damage after oxidative stress. *Exp. Neurol.* **159**, 309–318.

Devauchelle, V., and Chiocchia, G. (2004). [What place for DNA microarray in inflammatory diseases?]. *Rev. Med. Interne* **25**, 732–739.

Devauchelle, V., Marion, S., Cagnard, N., Mistou, S., Falgarone, G., Breban, M., Letourneur, F., Pitaval, A., Alibert, O., Lucchesi, C., Anract, P., Hamadouche, M., *et al.* (2004). DNA microarray allows molecular profiling of rheumatoid arthritis and identification of pathophysiological targets. *Genes Immun.* **5**, 597–608.

Devauchelle, V., Essabbani, A., De Pinieux, G., Germain, S., Tourneur, L., Mistou, S., Margottin-Goguet, F., Anract, P., Migaud, H., Le Nen, D., Lequerre, T., Saraux, A., *et al.* (2006). Characterization and functional consequences of underexpression of clusterin in rheumatoid arthritis. *J. Immunol.* **177**, 6471–6479.

Dhodapkar, K. M., Banerjee, D., Connolly, J., Kukreja, A., Matayeva, E., Veri, M. C., Ravetch, J. V., Steinman, R. M., and Dhodapkar, M. V. (2007). Selective blockade of the inhibitory Fcgamma receptor (FcgammaRIIB) in human dendritic cells and monocytes induces a type I interferon response program. *J. Exp. Med.* **204**, 1359–1369.

Diel, P., Smolnikar, K., Schulz, T., Laudenbach-Leschowski, U., Michna, H., and Vollmer, G. (2001). Phytoestrogens and carcinogenesis-differential effects of genistein in experimental models of normal and malignant rat endometrium. *Hum. Reprod.* **16**, 997–1006.

Doyle, H. A., and Mamula, M. J. (2001). Post-translational protein modifications in antigen recognition and autoimmunity. *Trends Immunol.* **22**, 443–449.

Ehrenstein, M. R., Evans, J. G., Singh, A., Moore, S., Warnes, G., Isenberg, D. A., and Mauri, C. (2004). Compromised function of regulatory T cells in rheumatoid arthritis and reversal by anti-TNFalpha therapy. *J. Exp. Med.* **200**, 277–285.

Feldmann, M., Brennan, F. M., and Maini, R. N. (1996). Role of cytokines in rheumatoid arthritis. *Annu. Rev. Immunol.* **14**, 397–440.

Figdor, C. G., van Kooyk, Y., and Adema, G. J. (2002). C-type lectin receptors on dendritic cells and Langerhans cells. *Nat. Rev. Immunol.* **2**, 77–84.

Finck, B. K., Chan, B., and Wofsy, D. (1994). Interleukin 6 promotes murine lupus in NZB/NZW F1 mice. *J. Clin. Invest.* **94**, 585–591.

Firestein, G. S., and Zvaifler, N. J. (1990). How important are T cells in chronic rheumatoid synovitis? *Arthritis Rheum.* **33**, 768–773.

Firestein, G. S., and Zvaifler, N. J. (2002). How important are T cells in chronic rheumatoid synovitis? II. T cell-independent mechanisms from beginning to end. *Arthritis Rheum.* **46**, 298–308.

French, L. E., Polla, L. L., Tschopp, J., and Schifferli, J. A. (1992). Membrane attack complex (MAC) deposits in skin are not always accompanied by S-protein and clusterin. *J. Invest. Dermatol.* **98**, 758–763.

Fujinami, M., Sato, K., Kashiwazaki, S., and Aotsuka, S. (1997). Comparable histological appearance of synovitis in seropositive and seronegative rheumatoid arthritis. *Clin. Exp. Rheumatol.* **15**, 11–17.

Furuzawa-Carballeda, J., Vargas-Rojas, M. I., and Cabral, A. R. (2007). Autoimmune inflammation from the Th17 perspective. *Autoimmun. Rev.* **6**, 169–175.

Ghiso, J., Matsubara, E., Koudinov, A., Choi-Miura, N. H., Tomita, M., Wisniewski, T., and Frangione, B. (1993). The cerebrospinal-fluid soluble form of Alzheimer's amyloid beta is complexed to SP-40,40 (apolipoprotein J), an inhibitor of the complement membrane-attack complex. *Biochem. J.* **293** (Pt 1), 27–30.

Girton, R. A., Sundin, D. P., and Rosenberg, M. E. (2002). Clusterin protects renal tubular epithelial cells from gentamicin-mediated cytotoxicity. *Am. J. Physiol. Renal Physiol.* **282**, F703–F709.

Goronzy, J. J., and Weyand, C. M. (2001). Thymic function and peripheral T-cell homeostasis in rheumatoid arthritis. *Trends Immunol.* **22**, 251–255.

Gottenberg, J. E., and Chiocchia, G. (2007). Dendritic cells and interferon-mediated autoimmunity. *Biochimie* **89**, 856–871.

Guermonprez, P., Valladeau, J., Zitvogel, L., Thery, C., and Amigorena, S. (2002). Antigen presentation and T cell stimulation by dendritic cells. *Annu. Rev. Immunol.* **20**, 621–667.

Hardardottir, I., Kunitake, S. T., Moser, A. H., Doerrler, W. T., Rapp, J. H., Grunfeld, C., and Feingold, K. R. (1994). Endotoxin and cytokines increase hepatic messenger RNA levels and serum concentrations of apolipoprotein J (clusterin) in Syrian hamsters. *J. Clin. Invest.* **94**, 1304–1309.

Herrada, A. A., Contreras, F. J., Tobar, J. A., Pacheco, R., and Kalergis, A. M. (2007). Immune complex-induced enhancement of bacterial antigen presentation requires Fcgamma receptor III expression on dendritic cells. *Proc. Natl. Acad. Sci. USA* **104**, 13402–13407.

Hogasen, K., Mollnes, T. E., Harboe, M., Gotze, O., Hammer, H. B., and Oppermann, M. (1995). Terminal complement pathway activation and low lysis inhibitors in rheumatoid arthritis synovial fluid. *J. Rheumatol.* **22**, 24–28.

Hopwood, B., Tsykin, A., Findlay, D. M., and Fazzalari, N. L. (2007). Microarray gene expression profiling of osteoarthritic bone suggests altered bone remodelling, WNT and transforming growth factor-beta/bone morphogenic protein signalling. *Arthritis Res. Ther.* **9**, R100.

Humphreys, D. T., Carver, J. A., Easterbrook-Smith, S. B., and Wilson, M. R. (1999). Clusterin has chaperone-like activity similar to that of small heat shock proteins. *J. Biol. Chem.* **274**, 6875–6881.

Jackson, J. K., Gleave, M. E., Gleave, J., and Burt, H. M. (2005). The inhibition of angiogenesis by antisense oligonucleotides to clusterin. *Angiogenesis* **8**, 229–238.

Jaen, O., Petit-Teixeira, E., Kirsten, H., Ahnert, P., Semerano, L., Pierlot, C., Cornelis, F., Boissier, M. C., and Falgarone, G. (2009). No evidence of major effects in several Toll-like receptor gene polymorphisms in rheumatoid arthritis. *Arthritis Res. Ther.* **11**, R5.

Janeway, C. A., Jr., Goodnow, C. C., and Medzhitov, R. (1996). Danger-pathogen on the premises! Immunological tolerance. *Curr. Biol.* **6**, 519–522.

Jenne, D. E., Lowin, B., Peitsch, M. C., Bottcher, A., Schmitz, G., and Tschopp, J. (1991). Clusterin (complement lysis inhibitor) forms a high density lipoprotein complex with apolipoprotein A-I in human plasma. *J. Biol. Chem.* **266**, 11030–11036.

Kang, Y. K., Hong, S. W., Lee, H., and Kim, W. H. (2004). Overexpression of clusterin in human hepatocellular carcinoma. *Hum. Pathol.* **35**, 1340–1346.

Kaisman-Elbaz, T., Sekler, I., Fishman, D., Karol, N., Forberg, M., Kahn, N., Hershfinkel, M., and Silverman, W. F. (2009). Cell death induced by zinc and cadmium is mediated by clusterin in cultured mouse seminiferous tubules. *J. Cell. Physiol.* **220**, 222–229.

Kelso, G. J., Stuart, W. D., Richter, R. J., Furlong, C. E., Jordan-Starck, T. C., and Harmony, J. A. (1994). Apolipoprotein J is associated with paraoxonase in human plasma. *Biochemistry* **33**, 832–839.

Khan, I. M., Salter, D. M., Bayliss, M. T., Thomson, B. M., and Archer, C. W. (2001). Expression of clusterin in the superficial zone of bovine articular cartilage. *Arthritis Rheum.* **44**, 1795–1799.

Kohem, C. L., Brezinschek, R. I., Wisbey, H., Tortorella, C., Lipsky, P. E., and Oppenheimer-Marks, N. (1996). Enrichment of differentiated CD45RBdim,CD27- memory T cells in the peripheral blood, synovial fluid, and synovial tissue of patients with rheumatoid arthritis. *Arthritis Rheum.* **39**, 844–854.

Kounnas, M. Z., Loukinova, E. B., Stefansson, S., Harmony, J. A., Brewer, B. H., Strickland, D. K., and Argraves, W. S. (1995). Identification of glycoprotein 330 as an endocytic receptor for apolipoprotein J/clusterin. *J. Biol. Chem.* **270**, 13070–13075.

Kuhn, K. A., Kulik, L., Tomooka, B., Braschler, K. J., Arend, W. P., Robinson, W. H., and Holers, V. M. (2006). Antibodies against citrullinated proteins enhance tissue injury in experimental autoimmune arthritis. *J. Clin. Invest.* **116**, 961–973.

Kumar, S., Connor, J. R., Dodds, R. A., Halsey, W., Van Horn, M., Mao, J., Sathe, G., Mui, P., Agarwal, P., Badger, A. M., Lee, J. C., Gowen, M., *et al.* (2001). Identification and initial characterization of 5000 expressed sequenced tags (ESTs) each from adult human normal and osteoarthritic cartilage cDNA libraries. *Osteoarthr. Cartil.* **9**, 641–653.

Laborde, E. A., Vanzulli, S., Beigier-Bompadre, M., Isturiz, M. A., Ruggiero, R. A., Fourcade, M. G., Catalan Pellet, A. C., Sozzani, S., and Vulcano, M. (2007). Immune complexes inhibit differentiation, maturation, and function of human monocyte-derived dendritic cells. *J. Immunol.* **179**, 673–681.

Leskov, K. S., Klokov, D. Y., Li, J., Kinsella, T. J., and Boothman, D. A. (2003). Synthesis and functional analyses of nuclear clusterin, a cell death protein. *J. Biol. Chem.* **278**, 11590–11600.

Lourda, M., Trougakos, I. P., and Gonos, E. S. (2007). Development of resistance to chemotherapeutic drugs in human osteosarcoma cell lines largely depends on up-regulation of Clusterin/Apolipoprotein J. *Int. J. Cancer* **120**, 611–622.

Lutzky, V., Hannawi, S., and Thomas, R. (2007). Cells of the synovium in rheumatoid arthritis. Dendritic cells. *Arthritis Res. Ther.* **9**, 219.

Makarov, S. S. (2001). NF-kappa B in rheumatoid arthritis: A pivotal regulator of inflammation, hyperplasia, and tissue destruction. *Arthritis Res.* **3**, 200–206.

Markopoulou, S., Kontargiris, E., Batsi, C., Tzavaras, T., Trougakos, I., Boothman, D. A., Gonos, E. S., and Kolettas, E. (2009). Vanadium-induced apoptosis of HaCaT cells is mediated by c-fos and involves nuclear accumulation of clusterin. *FEBS J.* **276**, 3784–3799.

Masson-Bessiere, C., Sebbag, M., Girbal-Neuhauser, E., Nogueira, L., Vincent, C., Senshu, T., and Serre, G. (2001). The major synovial targets of the rheumatoid arthritis-specific antifilaggrin autoantibodies are deiminated forms of the alpha- and beta-chains of fibrin. *J. Immunol.* **166**, 4177–4184.

Matsuda, A., Itoh, Y., Koshikawa, N., Akizawa, T., Yana, I., and Seiki, M. (2003). Clusterin, an abundant serum factor, is a possible negative regulator of MT6-MMP/MMP-25 produced by neutrophils. *J. Biol. Chem.* **278**, 36350–36357.

Matsumoto, I., Maccioni, M., Lee, D. M., Maurice, M., Simmons, B., Brenner, M., Mathis, D., and Benoist, C. (2002). How antibodies to a ubiquitous cytoplasmic enzyme may provoke joint-specific autoimmune disease. *Nat. Immunol.* **3**, 360–365.

Matzinger, P. (1994). Tolerance, danger, and the extended family. *Annu. Rev. Immunol.* **12**, 991–1045.

McHattie, S., Wells, G. A., Bee, J., and Edington, N. (1999). Clusterin in bovine spongiform encephalopathy (BSE). *J. Comp. Pathol.* **121**, 159–171.

McLaughlin, L., Zhu, G., Mistry, M., Ley-Ebert, C., Stuart, W. D., Florio, C. J., Groen, P. A., Witt, S. A., Kimball, T. R., Witte, D. P., Harmony, J. A., and Aronow, B. J. (2000). Apolipoprotein J/clusterin limits the severity of murine autoimmune myocarditis. *J. Clin. Invest.* **106**, 1105–1113.

Metz, T. O., Qian, W. J., Jacobs, J. M., Gritsenko, M. A., Moore, R. J., Polpitiya, A. D., Monroe, M. E., Camp, D. G., II, Mueller, P. W., and Smith, R. D. (2008). Application of proteomics in the discovery of candidate protein biomarkers in a diabetes autoantibody standardization program sample subset. *J. Proteome Res.* **7**, 698–707.

Miagkov, A. V., Kovalenko, D. V., Brown, C. E., Didsbury, J. R., Cogswell, J. P., Stimpson, S. A., Baldwin, A. S., and Makarov, S. S. (1998). NF-kappaB activation provides the potential link between inflammation and hyperplasia in the arthritic joint. *Proc. Natl. Acad. Sci. USA* **95**, 13859–13864.

Moll, S., Menoud, P. A., French, L., Sappino, A. P., Pastore, Y., Schifferli, J. A., and Izui, S. (1998). Tubular up-regulation of clusterin mRNA in murine lupus-like nephritis. *Am. J. Pathol.* **152**, 953–962.

Moretti, R. M., Marelli, M. M., Mai, S., Cariboni, A., Scaltriti, M., Bettuzzi, S., and Limonta, P. (2007). Clusterin isoforms differentially affect growth and motility of prostate cells: Possible implications in prostate tumorigenesis. *Cancer Res.* **67**, 10325–10333.

Murphy, B. F., Kirszbaum, L., Walker, I. D., and d'Apice, A. J. (1988). SP-40,40, a newly identified normal human serum protein found in the SC5b-9 complex of complement and in the immune deposits in glomerulonephritis. *J. Clin. Invest.* **81**, 1858–1864.

Newkirk, M. M., Apostolakos, P., Neville, C., and Fortin, P. R. (1999). Systemic lupus erythematosus, a disease associated with low levels of clusterin/apoJ, an antiinflammatory protein. *J. Rheumatol.* **26**, 597–603.

Nimmerjahn, F., and Ravetch, J. V. (2006). Fcgamma receptors: Old friends and new family members. *Immunity* **24**, 19–28.

Olsen, N. J., and Stein, C. M. (2004). New drugs for rheumatoid arthritis. *N. Engl. J. Med.* **350**, 2167–2179.

Page, G., Lebecque, S., and Miossec, P. (2002). Anatomic localization of immature and mature dendritic cells in an ectopic lymphoid organ: Correlation with selective chemokine expression in rheumatoid synovium. *J. Immunol.* **168**, 5333–5341.

Pucci, S., Paola, M., Fiabola, S., David, B. A., and Luigi, S. G. (2009). Interleukin-6 affects cell death escaping mechanisms acting on Bax–Ku70–Clusterin interactions in human colon cancer progression. *Cell Cycle* **8**, 473–481.

Radstake, T. R., Blom, A. B., Sloetjes, A. W., van Gorselen, E. O., Pesman, G. J., Engelen, L., Torensma, R., van den Berg, W. B., Figdor, C. G., van Lent, P. L., Adema, G. J., and Barrera, P. (2004). Increased FcgammaRII expression and aberrant tumour necrosis factor alpha production by mature dendritic cells from patients with active rheumatoid arthritis. *Ann. Rheum. Dis.* **63**, 1556–1563.

Reddy, K. B., Karode, M. C., Harmony, A. K., and Howe, P. H. (1996). Interaction of transforming growth factor beta receptors with apolipoprotein J/clusterin. *Biochemistry* **35**, 309–314.

Rizzi, F., Caccamo, A. E., Belloni, L., and Bettuzzi, S. (2009). Clusterin is a short half-life, polyubiquitinated protein, which controls the fate of prostate cancer cells. *J. Cell. Physiol.* **219**, 314–323.

Roeth, P. J., and Easterbrook-Smith, S. B. (1996). C1q is a nucleotide binding protein and is responsible for the ability of clusterin preparations to promote immune complex formation. *Biochim. Biophys. Acta* **1297**, 159–166.

Rosenberg, M. E., Girton, R., Finkel, D., Chmielewski, D., Barrie, A., III, Witte, D. P., Zhu, G., Bissler, J. J., Harmony, J. A., and Aronow, B. J. (2002). Apolipoprotein J/clusterin prevents a progressive glomerulopathy of aging. *Mol. Cell. Biol.* **22**, 1893–1902.

Rudd, P. M., Elliott, T., Cresswell, P., Wilson, I. A., and Dwek, R. A. (2001). Glycosylation and the immune system. *Science* **291**, 2370–2376.

Salmond, R. J., Filby, A., Qureshi, I., Caserta, S., and Zamoyska, R. (2009). T-cell receptor proximal signaling via the Src-family kinases, Lck and Fyn, influences T-cell activation, differentiation, and tolerance. *Immunol. Rev.* **228**, 9–22.

Santarlasci, V., Maggi, L., Capone, M., Frosali, F., Querci, V., De Palma, R., Liotta, F., Cosmi, L., Maggi, E., Romagnani, S., and Annunziato, F. (2009). TGF-beta indirectly favors the development of human Th17 cells by inhibiting Th1 cells. *Eur. J. Immunol.* **39**, 207–215.

Santilli, G., Aronow, B. J., and Sala, A. (2003). Essential requirement of apolipoprotein J (clusterin) signaling for IkappaB expression and regulation of NF-kappaB activity. *J. Biol. Chem.* **278**, 38214–38219.

Savkovic, V., Gantzer, H., Reiser, U., Selig, L., Gaiser, S., Sack, U., Kloppel, G., Mossner, J., Keim, V., Horn, F., and Bodeker, H. (2007). Clusterin is protective in pancreatitis through antiapoptotic and anti-inflammatory properties. *Biochem. Biophys. Res. Commun.* **356**, 431–437.

Scaltriti, M., Brausi, M., Amorosi, A., Caporali, A., D'Arca, D., Astancolle, S., Corti, A., and Bettuzzi, S. (2004a). Clusterin (SGP 2, ApoJ) expression is downregulated in low- and high-grade human prostate cancer. *Int. J. Cancer* **108**, 23–30.

Scaltriti, M., Santamaria, A., Paciucci, R., and Bettuzzi, S. (2004b). Intracellular clusterin induces G2-M phase arrest and cell death in PC-3 prostate cancer cells1. *Cancer Res.* **64**, 6174–6182.

Slawin, K., Sawczuk, I. S., Olsson, C. A., and Buttyan, R. (1990). Chromosomal assignment of the human homologue encoding SGP-2. *Biochem. Biophys. Res. Commun.* **172**, 160–164.

Steinman, R. M. (2007). Dendritic cells: Understanding immunogenicity. *Eur. J. Immunol.* **37** (Suppl. 1), S53–S60.

Steinman, R. M., and Banchereau, J. (2007). Taking dendritic cells into medicine. *Nature* **449**, 419–426.

Steinman, R. M., Hawiger, D., Liu, K., Bonifaz, L., Bonnyay, D., Mahnke, K., Iyoda, T., Ravetch, J., Dhodapkar, M., Inaba, K., and Nussenzweig, M. (2003). Dendritic cell function *in vivo* during the steady state: A role in peripheral tolerance. *Ann. N. Y. Acad. Sci.* **987**, 15–25.

Stevens, A. L., Wishnok, J. S., Chai, D. H., Grodzinsky, A. J., and Tannenbaum, S. R. (2008). A sodium dodecyl sulfate-polyacrylamide gel electrophoresis-liquid chromatography tandem mass spectrometry analysis of bovine cartilage tissue response to mechanical compression injury and the inflammatory cytokines tumor necrosis factor alpha and interleukin-1beta. *Arthritis Rheum.* **58**, 489–500.

Tak, P. P., and Firestein, G. S. (2001). NF-kappaB: A key role in inflammatory diseases. *J. Clin. Invest.* **107**, 7–11.

Tak, P. P., Gerlag, D. M., Aupperle, K. R., van de Geest, D. A., Overbeek, M., Bennett, B. L., Boyle, D. L., Manning, A. M., and Firestein, G. S. (2001). Inhibitor of nuclear factor kappaB kinase beta is a key regulator of synovial inflammation. *Arthritis Rheum.* **44**, 1897–1907.

Trougakos, I. P., So, A., Jansen, B., Gleave, M. E., and Gonos, E. S. (2004). Silencing expression of the clusterin/apolipoprotein j gene in human cancer cells using small interfering RNA induces spontaneous apoptosis, reduced growth ability, and cell sensitization to genotoxic and oxidative stress. *Cancer Res.* **64**, 1834–1842.

Trougakos, I. P., Lourda, M., Antonelou, M. H., Kletsas, D., Gorgoulis, V. G., Papassideri, I. S., Zou, Y., Margaritis, L. H., Boothman, D. A., and Gonos, E. S. (2009). Intracellular clusterin inhibits mitochondrial apoptosis by suppressing p53-activating stress signals and stabilizing the cytosolic Ku70–Bax protein complex. *Clin. Cancer Res.* **15**, 48–59.

Tschopp, J., Chonn, A., Hertig, S., and French, L. E. (1993). Clusterin, the human apolipoprotein and complement inhibitor, binds to complement C7, C8 beta, and the b domain of C9. *J. Immunol.* **151**, 2159–2165.

Tuncdemir, M., and Ozturk, M. (2008). The effects of ACE inhibitor and angiotensin receptor blocker on clusterin and apoptosis in the kidney tissue of streptozotocin-diabetic rats. *J. Mol. Histol.* **39**, 605–616.

van Gaalen, F. A., van Aken, J., Huizinga, T. W., Schreuder, G. M., Breedveld, F. C., Zanelli, E., van Venrooij, W. J., Verweij, C. L., Toes, R. E., and de Vries, R. R. (2004). Association between HLA class II genes and autoantibodies to cyclic citrullinated peptides (CCPs) influences the severity of rheumatoid arthritis. *Arthritis Rheum.* **50**, 2113–2121.

van Vliet, S. J., den Dunnen, J., Gringhuis, S. I., Geijtenbeek, T. B., and van Kooyk, Y. (2007). Innate signaling and regulation of dendritic cell immunity. *Curr. Opin. Immunol.* **19**, 435–440.

Vignali, D. A., Collison, L. W., and Workman, C. J. (2008). How regulatory T cells work. *Nat. Rev. Immunol.* **8**, 523–532.

Voo, K. S., Wang, Y. H., Santori, F. R., Boggiano, C., Wang, Y. H., Arima, K., Bover, L., Hanabuchi, S., Khalili, J., Marinova, E., Zheng, B., Littman, D. R., *et al.* (2009). Identification of IL-17-producing FOXP3+ regulatory T cells in humans. *Proc. Natl. Acad. Sci. USA* **106**, 4793–4798.

Wang, L., Erling, P., Bengtsson, A. A., Truedsson, L., Sturfelt, G., and Erlinge, D. (2004). Transcriptional down-regulation of the platelet ADP receptor P2Y(12) and clusterin in patients with systemic lupus erythematosus. *J. Thromb. Haemost.* **2**, 1436–1442.

Wells, A. D. (2009). New insights into the molecular basis of T cell anergy: Anergy factors, avoidance sensors, and epigenetic imprinting. *J. Immunol.* **182**, 7331–7341.

Weyand, C. M., Goronzy, J. J., Takemura, S., and Kurtin, P. J. (2000). Cell–cell interactions in synovitis. Interactions between T cells and B cells in rheumatoid arthritis. *Arthritis Res.* **2**, 457–463.

Wilson, M. R., and Easterbrook-Smith, S. B. (1992). Clusterin binds by a multivalent mechanism to the Fc and Fab regions of IgG. *Biochim. Biophys. Acta* **1159**, 319–326.

Wilson, N. J., Boniface, K., Chan, J. R., McKenzie, B. S., Blumenschein, W. M., Mattson, J. D., Basham, B., Smith, K., Chen, T., Morel, F., Lecron, J. C., Kastelein, R. A., et al. (2007). Development, cytokine profile and function of human interleukin 17-producing helper T cells. *Nat. Immunol.* **8**, 950–957.

Yang, C. R., Leskov, K., Hosley-Eberlein, K., Criswell, T., Pink, J. J., Kinsella, T. J., and Boothman, D. A. (2000). Nuclear clusterin/XIP8, an x-ray-induced Ku70-binding protein that signals cell death. *Proc. Natl. Acad. Sci. USA* **97**, 5907–5912.

Zhang, H., Kim, J. K., Edwards, C. A., Xu, Z., Taichman, R., and Wang, C. Y. (2005). Clusterin inhibits apoptosis by interacting with activated Bax. *Nat. Cell Biol.* **7**, 909–915.

Oxidative Stress in Malignant Progression: The Role of Clusterin, A Sensitive Cellular Biosensor of Free Radicals

Ioannis P. Trougakos* and Efstathios S. Gonos[†]

*Department of Cell Biology and Biophysics, Faculty of Biology, University of Athens, Panepistimiopolis Zografou, Athens 15784, Greece
[†]Institute of Biological Research and Biotechnology, National Hellenic Research Foundation, Athens 11635, Greece

Clusterin/Apolipoprotein J (CLU) gene is expressed in most human tissues and encodes for two protein isoforms; a conventional heterodimeric secreted glycoprotein and a truncated nuclear form. CLU has been functionally implicated in several physiological processes as well as in many pathological conditions including ageing, diabetes, atherosclerosis, degenerative diseases, and tumorigenesis. A major link of all these, otherwise unrelated, diseases is that they are characterized by increased oxidative injury due to impaired balance between production and disposal of reactive oxygen or nitrogen species. Besides the aforementioned diseases, CLU gene is differentially regulated by a wide variety of stimuli which may also promote the production of reactive species including cytokines, interleukins, growth factors, heat shock, radiation, oxidants, and chemotherapeutic drugs. Although at low concentration reactive species may contribute to normal cell signaling and homeostasis, at increased amounts they promote genomic instability, chronic inflammation, lipid oxidation, and amorphous aggregation of target proteins predisposing thus cells for carcinogenesis or other age-related disorders. CLU seems to intervene to these processes due to its small heat-shock protein-like chaperone activity being demonstrated by its property to inhibit protein aggregation and precipitation, a main feature of oxidant injury. The combined presence of many potential

Advances in CANCER RESEARCH
Copyright 2009, Elsevier Inc. All rights reserved.

0065-230X/09 $35.00
DOI: 10.1016/S0065-230X(09)04009-3

regulatory elements in the CLU gene promoter, including a Heat-Shock Transcription Factor-1 and an Activator Protein-1 element, indicates that CLU gene is an extremely sensitive cellular biosensor of even minute alterations in the cellular oxidative load. This review focuses on CLU regulation by oxidative injury that is the common molecular link of most, if not all, pathological conditions where CLU has been functionally implicated.

I. INTRODUCTION

The conventional isoform of the human Clusterin (CLU) gene (NCBI: NM_203339.1/NP_976084.1, CLU isoform 2) encodes for a ubiquitously expressed secreted protein of $\sim$75 kDa (sCLU) (Wong et al., 1994) being involved in several physiological processes including lipid transportation in the serum (de Silva et al., 1990), development, and differentiation (Aronow et al., 1993; Jenne and Tschopp, 1989; Trougakos et al., 2006b) (see also chapter "Cell Protective Functions of Secretory CLU"). In certain cell types, apoptosis induction correlates with the appearance of alternative CLU protein isoforms which mostly localize in cell nucleus (nCLU). nCLU can be produced by either alternative splicing or internal translation initiation; in both cases, the N-terminal sequence of sCLU that encodes the hydrophobic endoplasmic reticulum (ER) targeting peptide is bypassed resulting in cytosolic or nuclear protein localization (Leskov et al., 2003; Yang et al., 2000). Almost invariably nCLU exerts a cytostatic and/or proapoptotic function (Leskov et al., 2003; Moretti et al., 2007; Yang et al., 2000) (see also chapter "Nuclear CLU (nCLU) and the Fate of the Cell").

Two main functions have been assigned to sCLU up to date. The one refers to its Apolipoprotein function at the high-density lipoprotein (HDL) particles (de Silva et al., 1990) and the second to its ability to bind to hydrophobic regions of partially unfolded proteins and via an ATP-independent mechanism to inhibit protein aggregation and precipitation otherwise caused by physical or chemical stresses (e.g., heat, reduction) (Humphreys et al., 1999; Poon et al., 2000). On the basis of this latter property, sCLU was classified as an extracellular functional homologue of the small heat-shock proteins (sHSPs) (Wilson and Easterbrook-Smith, 2000) (see also chapter "The Chaperone Action of CLU and Its Putative Role in Quality Control of Extracellular Protein Folding"). Interestingly, it seems that in certain cell types sCLU can escape the secretory pathway and localize in the nucleocytosolic continuum and mitochondria (Nizard et al., 2007; Trougakos et al., 2009b; Zhang et al., 2005) indicating that sCLU may also function as an intracellular chaperone.

The CLU gene promoter is highly conserved and contains several regulatory elements that regulate the complex tissue-specific control of the gene.

Among these elements are a conserved 14-bp "CLU-specific element" (CLE), which is recognized by the Heat-Shock Transcription Factor 1 (HSF1) and a site for the Activator Protein-1 (AP-1) (Jin and Howe, 1997; Michel *et al.*, 1995; Wong *et al.*, 1994). By acting in synergy, the CLE and AP-1 elements can make the CLU gene promoter particularly sensitive to even minute environmental changes or alterations in the cellular oxidative load. In support, the CLU gene is responsive to a wide variety of both intrinsic or extrinsic factors including oncogenes, transcription and growth factors, cytokines and most stress-inducing agents including oxidants, hyperoxia, proteotoxic stress, heavy metals, UVA, UVB, ionizing radiation (IR), heat shock, and chemotherapeutic drugs (Calero *et al.*, 2005; Klokov *et al.*, 2004; Trougakos and Gonos, 2002, 2006) (see also chapter "Regulation of CLU Gene Expression by Oncogenes, Epigenetic Regulation, and the Role of CLU in Tumorigenesis," Vol. 105). Moreover, CLU is differentially regulated and have been functionally involved in a wide range of distinct pathological conditions including atherosclerosis, diabetes, kidney, and neuron degeneration as well as carcinogenesis (Trougakos and Gonos, 2006); in most, if not all, of these pathologies differential regulation referred to sCLU.

The only common characteristic shared by all these, otherwise unrelated in their etiology and/or clinical manifestation, pathological conditions is the fact that they are characterized by increased oxidative stress and injury, which may then promote proteotoxic stress, genomic instability, arrest of proliferation, and high rates of cellular death. In the current review, we promote the idea that sCLU is a sensitive cellular biosensor of oxidative stress that functions to protect cells from the deleterious effects of free radicals and their derivatives.

II. MOLECULAR EFFECTS OF ALTERED OXIDATIVE LOAD IN HUMAN CELLS

Oxygen is essential to support aerobic metabolism and life. However, since the normal cellular milieu is reductive, oxygen molecules undergo univalent reductions forming oxygen-free radicals (Beckman and Ames, 1998). In the last two decades there has been an significant interest in the role of oxygen-free radicals, more generally known as "reactive oxygen species" (ROS) and "reactive nitrogen species" (RNS) in experimental and clinical medicine (Dröge, 2003). Oxygen-free radicals can be produced by both endogenous and exogenous sources. Endogenously oxygen-free radicals may arise as by-products of normal aerobic metabolism (i.e., excessive stimulation of NAD(P)H oxidases, oxidative energy metabolism, P450 metabolism, and peroxisomes), as second messengers in various signal transduction pathways or during inflammatory neutrophils and macrophage cells activation. Exogenous sources of free radical production are various types of radiations (e.g., UV light, X-rays, or gamma rays), metal-catalyzed reactions, and atmospheric pollutants (e.g., chlorinated compounds) (Dröge, 2002; Martindale and Holbrook, 2002; Valko et al., 2006).

Oxidative stress occurs due to overproduction of oxygen-free radicals that exceeds the cell's antioxidant capacity (Klaunig and Kamendulis, 2004). Oxygen-free radicals, whether produced endogenously as a consequence of normal cell functions or derived from external sources, pose a constant threat to cells living in an aerobic environment as they can severely damage DNA, proteins, and lipids (Martindale and Holbrook, 2002). More that 100 oxidative DNA adducts have been identified (Demple and Harrison, 1994) and in a given cell, an estimated 10^5 oxidative lesions per day are formed (Fraga et al., 1990). Cells contain a number of antioxidant defenses (either nonenzymatic antioxidants or antioxidant enzymes) to minimize fluctuations in oxygen-free radicals. Nonenzymatic antioxidants include vitamins E and C, natural flavonoids, carotenoids, thiol antioxidants (glutathione, thioredoxin, and lipoic acid), melatonin, and other compounds (McCall and Frei, 1999), while enzymatic antioxidants include catalase, glutathione peroxidases, and superoxide dismutase (SOD) (Mates et al., 1999). Host survival depends upon the ability of cells and tissues to adapt to or resist the stress, and repair or remove damaged molecules or cells. Numerous stress response mechanisms have evolved for these purposes and they are rapidly activated in response to oxidative insults (Valko et al., 2006).

Oxygen-free radicals are well recognized for playing a dual role as both beneficial and deleterious species. More specifically, low physiological amounts of oxidants may act as a secondary messenger in intracellular signaling cascades promoting normal cell homeostasis and proliferation. Abnormally high levels of ROS may, however, result in deregulation of redox-sensitive signaling pathways and extensive cell damage leading to responses which may include arrest or induction of transcription, induction of signal transduction pathways, replication errors, and genomic instability, reversible growth arrest, senescence, or cellular death (Fig. 1) (Storz, 2005; Valko et al., 2006).

Cellular damage due to increased oxidative stress may be reflected to DNA, proteins, and lipids. DNA damage is mainly induced via the formation of 8-hydroxy-2'-deoxyguanosine (8-oxodG) that is a hydroxyl radical-damaged guanine nucleotide. Additional effects may relate to single- or double-stranded DNA breaks, purine, pyrimidine, or deoxyribose modifications, and DNA cross-links (Floyd et al., 1990). Protein modifications relate to direct protein oxidation (e.g., at cysteine residues), formation of peroxyl radicals (Stadtman, 1992), conjugation with lipid peroxidation products, glycation, and glycoxidation (Chondrogianni and Gonos, 2005;

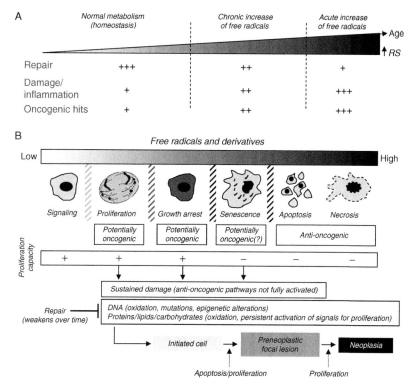

Fig. 1 Oxidative stress and cancer. (A) During ageing the load of oxidants in the organism gradually increases, due to both intrinsic and extrinsic factors, resulting in the accumulation of damaged molecules (e.g., nucleic acids, proteins, lipids, etc.), the impairment of the cell signaling regulatory pathways, and consequently the increase of the oncogenic hits. Thus, these effects directly relate to age-related diseases including cancer. The process can be significantly accelerated by various factors including several metabolic disorders (e.g., diabetes, obesity, etc.) or unhealthy habits (e.g., smoking, prolonged exposure to sun, etc.). (B) At the cellular level free radicals are not always deleterious as at low nonhazardous amounts contribute to physiological signaling and homeostasis. However, as the intracellular and/or extracellular amounts of free radicals and their derivatives increase so does the damage resulting in various responses which may include growth arrest, premature senescence, apoptosis, or necrosis. Although some are clearly tumor suppressing states (e.g., apoptosis and necrosis) others maybe potentially oncogenic as they combine gradual damage accumulation with a retained ability for proliferation. As the repair mechanisms weaken over time (because they are also targets of oxidative damage) more damaged molecules accumulate. This array of events increases dramatically the chances for the appearance of tumor initiating cells which may then give rise to preneoplastic focal lesions and eventually to neoplasia. CLU has been functionally implicated in all cellular states or responses described. (See Page 7 in Color Section at the back of the book.)

Poppek and Grune, 2006). These modifications may lead to loss of function, gain of function, or switch to a different function. Finally, lipid alterations mainly refer to lipid peroxidation (Yu and Chung, 2006).

Oxidative stress regulates an immense array of distinct signaling pathways that conclude in the activation of several transcription factors including, AP-1, nuclear factor-kappa beta (NF-κB), HSF1, p53, and signal transducer and activator of transcription (STAT) (Martindale and Holbrook, 2002; Storz, 2005). AP-1 is a collection of dimeric basic region-leucine zipper proteins that belong to the Jun, Fos, Maf, and ATF subfamilies (Chinenov and Kerppola, 2001). AP-1 activity is induced by H_2O_2 as well as from several cytokines and other physical and chemical stresses (Klatt et al., 1999). Activation of AP-1 mostly results in increased cell proliferation (Shaulian and Karin, 2001), while AP-1 proteins may also participate in oncogenic transformation by interacting with activated oncogenes such as H-Ras (Schutte et al., 1989). NF-κB is a ubiquitously expressed inducible transcription factor which in its active form is a dimer consisting of proteins from the Rel family, namely p50, p52, c-Rel, v-Rel, p65, and Rel B (Chen et al., 2001). NF-κB regulates a wide array of different genes involved in immune function, inflammation, stress response, differentiation, apoptosis, cell survival, and growth. Involvement of NF-κB in tumor initiation and progression mostly occurs in tissues in which cancer-related inflammation typically occurs (such as the gastrointestinal tract and the liver), while recent studies have shown that NF-κB signaling may also play critical roles in the epithelial-to-mesenchymal transition (EMT) and cancer stem cells physiology (Sarkar et al., 2008). Activated NF-κB, functions together with other transcription factors, such as AP-1 to regulate the expression of immune and inflammatory genes including interleukins (IL)-2, -2R, -6, and -8 and tumor necrosis factor-alpha (TNF-α). Most of these proteins can also cause the activation of NF-κB, forming a positive regulatory loop to amplify inflammatory responses (Chen et al., 2001; Sarkar et al., 2008). HSF1, under normal conditions is kept inactive in the cytosol by mainly binding to HSP90 (Morimoto, 2002). Upon stress, the suppressing chaperones become occupied by the misfolded proteins liberating thus HSF1 which can then translocate to the nucleus and activate several target genes (Morimoto, 2002; Soti et al., 2005). P53, the so-called "genome guardian" (Lane, 1992), is critically involved in the processes of many different stress responses and it is disrupted in more than half of all human cancers (Vousden and Lane, 2007). Besides the aforementioned nuclear transcription factors, free radicals can also regulate a wide panel of receptor tyrosine kinases, nonreceptor protein kinases, the H-Ras oncogene, protein tyrosine phosphatases, and serine/threonine kinases [e.g., transforming growth factor-beta (TGF-β)] (Valko et al., 2006).

III. OXIDATIVE INJURY IN AGEING AND AGE-RELATED DISEASES

Ageing is an inevitable consequence of life for nearly all organisms. It mostly reflects the outcome of complicated interactions between genetic factors along with the accumulation of a variety of deleterious stochastic changes over time due to a progressive failure of homeostasis that promotes multiple biochemical, molecular, and cellular changes which lead to increased disability, morbidity, and inevitably to death (Kirkwood, 2002). Thus, ageing is an adaptive process, not caused by a single factor or process, but it is rather a multifactorial procedure modulated by the interplay among genetic and environmental factors (Lane, 2003).

According to the "free radical theory of ageing" because of their high reactivity, oxygen-free radicals would lead to unavoidable and potentially deleterious by-products. Such an increasingly damaging process could be responsible for degenerative diseases and ageing (Beckman and Ames, 1998). Thus, it is generally accepted that oxidative stress is a major risk in ageing (Finkel and Holbrook, 2000; Golden *et al.*, 2002). In support, increased levels of oxidative and proteotoxic stress, DNA damage along with progressive changes in free radical-mediated regulatory processes have been linked to *in vivo* ageing in numerous reports (Carrard *et al.*, 2002; Grune *et al.*, 2004; Mariani *et al.*, 2005).

In invertebrates, such as *Caenorhabditis elegans* and *Drosophila melanogaster*, several mutants which reduce damage by oxidants result in increased life span providing thus a correlation between oxidative stress, life span, and genetic components (Hekimi and Guarente, 2003). Similarly, in mammals, mutation of the p66^{SHC} protein increased resistance to oxidative stress and prolonged life span (Migliaccio *et al.*, 1999). The level of p66^{SHC} decreased with advanced age (Sagi *et al.*, 2005) and ROS-mediated dysfunction in aortic endothelial cells was prevented in p66$^{SHC-/-}$ mice (Francia *et al.*, 2004). ROS implication in human ageing is also supported by *in vitro* studies at various cellular models. Specifically, normal human cells undergo a limited number of divisions in culture and progressively reach a state of irreversible growth arrest, a process termed as replicative senescence (RS). RS occurs because human cells (due to the absence of telomerase activity and owing to the biochemistry of DNA replication) acquire one or more critically short telomere (Campisi, 2005). Interestingly, cells can also senesce prematurely having long telomeres if exposed to a wide range of distinct types of stress including oxidative stress, proteotoxic stress, DNA damage, cytotoxic drugs, strong mitogenic signals, and disrupted chromatin (Collado and Serrano, 2006). This process is called stress-induced premature senescence (SIPS) and is phenotypically and biochemically similar to RS. Independently of the triggering event, *in vitro* senescing cells are characterized by accumulation of ROS, reduced proteasome activity, and high rates of protein oxidation/-carbonylation (Chondrogianni and Gonos, 2005; Trougakos *et al.*, 2002a).

Parallel to its continuous effects during life span there is growing evidence implicating oxidative damage to DNA, proteins, and lipids in the pathogenesis of various diseases. More specifically, oxidative stress plays a central role in various pathological conditions such as atherosclerosis, diabetes, neurodegeneration, inflammation, rheumatoid arthritis, and cancer (Dröge, 2002; Grune *et al.*, 2004; Mariani *et al.*, 2005; Sarkar *et al.*, 2008).

Increased production of ROS has been related to the risk for cardiovascular diseases such as atherosclerosis, angina, and myocardial infarction (Yoshizumi *et al.*, 2001). In fact, atherosclerosis represents a state of increased oxidative stress characterized by early lipid and protein oxidation in the vascular wall either at cardiovascular or cerebrovascular diseases (Stocker and Keaney, 2004). In support, heart failure associates with oxidative stress in animal studies, while during congestive heart failure the activation of most of the neurohormonal and inflammatory signaling pathways induce oxidative stress through several mechanisms including angiotensin II, aldosterone, TNF-α, and proinflammatory cytokines (Hill and Singal, 1997; Li *et al.*, 1998). ROS accumulation in pancreas is believed to play a central role in beta cells death and type I diabetes progression which is characterized by the severe destruction of insulin-producing beta cells (Ho and Bray, 1999). Most signal transduction responses due to pancreas-specific ROS production are mediated by activated NF-κB (Mollah *et al.*, 2008). Brain aging is associated with a progressive imbalance between antioxidant defenses and intracellular concentrations of oxygen-free radicals as exemplified by increases in products of DNA, protein, and lipid oxidation (Dröge and Schipper, 2007). Several lines of evidence highlight ROS implication in neurodegenerative diseases. Both brain ischemia and the condition of ischemia/reperfusion occurring after stroke, has been associated with free radical-mediated reactions (Alexandrova *et al.*, 2004) while most studies agree that oxidative stress contributes to dopaminergic cell degeneration in Parkinson's disease (Jenner, 2003) and that oxidative stress is one of the earliest events in Alzheimer's disease (AD) (Zhu *et al.*, 2004). Administration of vitamin E leads to a slowing of the AD progression and the patients taking antioxidant vitamins and anti-inflammatory compounds have a lower incidence of AD (Mancuso *et al.*, 2007). Oxidizing conditions cause protein cross-linking and aggregation of Aβ peptides (Dyrks *et al.*, 1993) and also contribute to aggregation of tau and other cytoskeletal proteins (Bellomo and Mirabelli, 1992). Finally, ROS seem to contribute significantly to tissue injury and inflammation in rheumatoid arthritis and the control of inflammation in arthritic patients by antioxidants could become an efficient strategy for antirheumatic therapy (Bauerova and Bezek, 1999).

IV. OXIDATIVE STRESS IN MALIGNANT PROGRESSION

Cancer development is a multifactorial, multistage process characterized by the cumulative action of additive events occurring in a single cell. It is generally described by three stages: initiation, promotion, and progression. Mutations, oncogene activation, genomic instability, and epigenetic factors deregulate gene expression profiles and disrupt the molecular networks providing tumor cells with sufficient diversity and proliferative advantage to evolve in the healthy tissue and eventually to metastasize (Halazonetis *et al.*, 2008; Weinberg, 2008).

Oxygen-free radicals are potential carcinogens because they can facilitate all three stages of tumor formation including mutagenesis, tumor promotion, and progression (Dröge, 2002; Halliwell, 2007; Klaunig and Kamendulis, 2004; Valko *et al.*, 2006). During initiation, a nonlethal mutation in DNA is followed by at least one round of DNA synthesis to fix the damage (e.g., due to 8-oxodG) and produce an altered cell. The promotion stage requires the continuous presence of the tumor promotion stimulus and therefore it is a reversible process. This stage may eventually result in the formation of an identifiable focal lesion which is characterized by the clonal expansion of finding cells via the induction of cell proliferation and/or inhibition of apoptosis. Production of ROS during this stage of carcinogenesis is the main line of ROS-related protumor effect as many carcinogens have a strong inhibiting effect on cellular antioxidant defense systems such as SOD, catalase, and glutathione. Low sustained level of oxidative stress in this state may also stimulate cell division resulting thus in tumor growth (Dreher and Junod, 1996). The final stage of the carcinogenic process is progression. This stage involves cellular and molecular changes that occur from the preneoplastic to the neoplastic state and is irreversible as it includes accumulation of additional genetic damage, genetic instability, and angiogenesis (Carmeliet, 2000).

Increased levels of oxidative DNA lesions (8-oxodG) have been noted in various tumors, strongly implicating ROS involvement in the etiology of cancer (Loft and Poulsen, 1996), while deficiency of various antioxidant defense enzymes raises oxidative damage levels and promotes age-related cancer development in animals (Elchuri *et al.*, 2005). The interaction of ROS with DNA has been reported to result in both fragmentation and strand breaks, as well as in decreased DNA methylation (Weitzman *et al.*, 1994). Additional evidence has implicated oxygen-free radicals and mitochondrial oxidative DNA damage in the carcinogenesis process (Cavalli and Liang, 1998). Aside from oxidized nucleic acids, other oxidation-derived DNA adducts that appear important in chemical carcinogenesis are those produced due to damage in cellular biomembranes that result in lipid peroxidation, a process that generates a variety of mutagenic products (Janero, 1990). As mentioned, cell exposure to nontoxic concentrations of ROS may result in aberrant increase of cell proliferation, survival, and cellular migration. These effects mostly relate to activation of AP-1 and NF-κB signal transduction pathways (Valko et al., 2006). The induction of oxidative stress modulates the function of several cell-cycle checkpoints (Shackelford et al., 2000) and may also inhibit intercellular communication (Upham et al., 1997) allowing the preneoplastic cell to escape the growth control of normal surrounding cells. Finally, recent data showed that the tumor suppressor genes p15^{INK4B} and p16^{INK4A} are among the major target genes in oxidative stress-induced carcinogenesis (Toyokuni, 2008). Therefore, there may be "fragile" sites in the genome that are more susceptible to oxidative stress.

Beyond the effects of oxygen-free radicals in tumorigenesis, the cancer cells are under increased and persistent oxidative stress. This elevated intrinsic oxygen-free radicals-related stress in cancer cells mostly relate to oncogenic stimulation, increased metabolic activity, mitochondrial malfunction, and low levels of enzymes such as SOD, glutathione peroxidase, glutathione reductase, and catalase (Gupte and Mumper, 2009). High endogenous levels of oxygen-free radicals have been found in human colorectal carcinoma (Kondo et al., 1999), in various types of leukemia (Devi et al., 2000) as well as in breast (Yeh et al., 2005), ovarian (Senthil et al., 2004), and stomach cancer (Batchioglu et al., 2006). Moreover, ROS levels in cancer cells relate to tumor aggressiveness as it has been reported that the difference in the redox status of two prostrate cancer cell lines correlated with the degree of their aggressiveness (Chaiswing et al., 2007).

Inflammatory process induces oxidative stress and reduces cellular antioxidant capacity. Chronic inflammation is a pathological condition characterized by continued active inflammation response and tissue destruction. In some types of cancer, inflammatory conditions are present before a malignant change occurs, while in others, an oncogenic change induces an inflammatory microenvironment that accelerates the development of tumors. The triggers of chronic inflammation that may increase cancer risk include microbial infections (i.e., Helicobacter pylori for gastric cancer and mucosal lymphoma), autoimmune diseases (i.e., inflammatory bowel disease for colon cancer), and cryptogenic inflammatory conditions (i.e., prostatitis for prostate cancer) (Mantovani et al., 2008). Particularly, regarding prostate cancer, there is emerging evidence that prostatic inflammation may contribute to prostate growth by introducing hyperplastic or neoplastic changes (Minelli et al., 2009; Sciarra et al., 2008). Pancreatic inflammation, mediated by cytokines, ROS, and upregulated proinflammatory pathways, may play a key role in the early development of pancreatic malignancy (Farrow and Evers, 2002), while epidemiological data have consistently demonstrated a positive relation between obesity and colorectal malignancy (Gunter and Leitzmann, 2006). Moreover, B lymphocytes are required for establishing chronic inflammatory states that promote *de novo* epithelial carcinogenesis (de Visser *et al.*, 2005).

The main substances that link inflammation to cancer via oxidative/nitro-sative stress are key intrinsic agents such as transcription factors (e.g., NF-κB and STAT3) and major inflammatory cytokines (e.g., IL-1β, IL-6, IL-23, and TNF-α) which lie downstream of NF-κB, the "master" gene transcription factor for promoting inflammation (Sarkar *et al.*, 2008). Further, the inflam-matory mediators such as cytokines, prostaglandins, and growth factors, which are regulated by NF-κB, can induce genetic and epigenetic changes including mutations in tumor suppressor genes, DNA methylation, and post-translational modifications, leading to the development and progression of cancer (Federico *et al.*, 2007).

On the basis of these facts it was recently proposed that inflammation may constitute the seventh hallmark that collectively dictate malignant growth (Mantovani, 2009); the other six being, self-sufficiency in growth signals, insensitivity to growth-inhibitory signals, evasion of programmed cell death, limitless replicative potential, sustained angiogenesis, and tissue invasion and metastasis (Hanahan and Weinberg, 2000).

V. CLUSTERIN AS A SENSITIVE CELLULAR BIOSENSOR OF OXIDATIVE STRESS

A. Clusterin Regulation and Function at the Cellular Level as a Response to Free Radicals and Their Derivatives

Considering that oxidants and oxidation injury modulate the activity of both the AP-1 and HSF1 transcription factors (Martindale and Holbrook, 2002) the combinatorial presence of both the AP-1 and CLE regulatory elements in the CLU gene promoter (Jin and Howe, 1997; Michel *et al.*, 1995) should make CLU gene an extremely sensitive biosensor to environ-mental changes and particularly to the direct or indirect downstream effects of free radicals and their derivatives.

Indeed, both sCLU (at low nontoxic doses) and nCLU (at high doses) are induced in human cells by IR, a potent inducer of ROS accumulation and genomic instability (Criswell *et al.*, 2005; Leskov *et al.*, 2003). Moreover, photooxidative cell damage in human cultured retinal pigmented epithelium upregulated nCLU (Ricci *et al.*, 2007). Proteotoxic stress induction due to proteasome inhibition results in sCLU mRNA upregulation via HSF1 and

HSF2 binding to the CLE element (Balantinou *et al.*, 2009; Loison *et al.*, 2006). Many oxidants (e.g., ethanol, tert-butylhydroperoxide, or H_2O_2) induce an immediate sCLU mRNA and protein upregulation in human diploid fibroblasts (Frippiat *et al.*, 2001), while CLU was found to be induced in retinal pigment epithelial cells after subtoxic oxidative stress via TGF-β release (Yu *et al.*, 2009) and in human neuroblastoma cells under conditions of increased production of ROS and lipid peroxidation (Strocchi *et al.*, 2006). At *in vivo* models both the organochlorine pesticides Methoxychlor and Lindane when administered in rats promoted a sequential reduction in the activities of catalase and SOD with concomitant increase in the levels of lipid peroxidation, HSP70, and sCLU (Saradha *et al.*, 2008; Vaithinathan *et al.*, 2009). A direct correlation of sCLU expression levels and organismal oxidative stress in also evident by the findings that sCLU concentration in human serum correlated positively with exposure to acute *in vivo* oxidative stress induced by heavy metals (Alexopoulos *et al.*, 2008), while, on the other hand, adult-onset of caloric restriction that retarded age-related oxidative damages resulted in CLU suppression in kidneys of rats (Chen *et al.*, 2008c). Finally, immunohistochemical analysis showed that CLU is constantly associated with altered elastic fibers in aged human skin and elastotic material of sun-damaged skin. It was found that CLU binds elastin, and effectively inhibited UV-induced aggregation of elastin by exerting its chaperone activity(Janig *et al.*, 2007).

Several studies demonstrate that sCLU suppress the deleterious effects of oxidants. More specifically, sCLU protects human epidermoid cancer cells from H_2O_2, superoxide anion, hyperoxia, and UVA (Schwochau *et al.*, 1998; Viard *et al.*, 1999); normal human fibroblasts from cytotoxicity induced by ethanol, tert-butylhydroperoxide, or UVB (Dumont *et al.*, 2002); human prostate cancer cells from oxidative stress-induced DNA damage (Miyake *et al.*, 2004); lung fibroblasts from cigarette smoke oxidants (Carnevali *et al.*, 2006); and human retinal cells from free-radical damage (Kim *et al.*, 2007a) or *in vitro* ischemia (Kim *et al.*, 2007b). Taken together these data clearly indicate the sensitivity of CLU gene promoter to alterations in the cell oxidative load and the cytoprotective role of sCLU in oxidative injury.

As it is shown in Fig. 1, oxidative stress in normal human cells can trigger a wide range of responses including proliferation, reversible growth arrest, senescence, or cellular death. Interestingly, CLU has been implicated in all these cellular states and processes.

More specifically, in certain cellular contexts sCLU or nCLU transgene-mediated overexpression has been reported to induce growth retardation at the G_2/M checkpoint, reduce DNA synthesis, and augment the cytostatic effect of genotoxic stress (Bettuzzi *et al.*, 2002; Scaltriti *et al.*, 2004b; Thomas-Tikhonenko *et al.*, 2004; Trougakos *et al.*, 2005) while extracellular CLU blocked cellular proliferation in prostate cancer cells (Zhou *et al.*, 2002). However, although high CLU levels may reduce the cycling potential of human cells for a short period of time, these cells eventually adapt to high CLU levels and continue cycling (Bettuzzi *et al.*, 2002; Trougakos *et al.*, 2005). Interestingly, exogenous sCLU promoted growth in astrocytes by activating epidermal growth factor receptor and through the Ras/Raf-1/MEK/ERK signaling cascade (Shim *et al.*, 2009). sCLU is a faithful biomarker of cellular senescence as it is upregulated during RS (Petropoulou *et al.*, 2001) or during SIPS induction by UVB (Debacq-Chainiaux *et al.*, 2005), subcytotoxic doses of H_2O_2 (Dumont *et al.*, 2000; Frippiat *et al.*, 2001), proteotoxic stress (Balantinou *et al.*, 2009), or cell exposure to heavy metals (Katsiki *et al.*, 2004).

nCLU has been mostly correlated to a pro-death function that in certain cell types is mediated via nuclear localization and Ku70 binding after cell exposure to IR (Leskov *et al.*, 2003; Yang *et al.*, 2000) or due to its ability to dismantle the actin cytoskeleton (Moretti *et al.*, 2007) (see also chapter "Nuclear CLU (nCLU) and the Fate of the Cell"). On the other hand, there is some controversy about sCLU role during cell death execution mainly due to opposing reported functions (Trougakos and Gonos, 2002). More specifically, in certain cellular contexts sCLU may enhance apoptosis after cell exposure to stress- or apoptosis-inducing agents (e.g., photodynamic therapy, chemotherapy, and oxidants) (Chen *et al.*, 2004; Kalka *et al.*, 2000; Scaltriti *et al.*, 2004b; Trougakos *et al.*, 2005) and high amounts of CLU may become cytotoxic when accumulated intracellularly (Debure *et al.*, 2003; Trougakos *et al.*, 2005). In HeLa cells growing under constraint conditions, CLU inhibits the phosphatidylinositol-3′-kinase–Akt pathway through attenuation of insulin-like growth factor 1 (Jo *et al.*, 2008). Also, in CLU knockout ($CLU^{-/-}$) mice it was found that CLU may exacerbate neuronal damage in hypoxia-ischemia (Han *et al.*, 2001). Contrary to these findings, which may relate to cell overload with high amounts of sCLU, in the majority of reported studies transgene-mediated sCLU overexpression potently protected mammalian cells from a variety of apoptosis-inducing agents including various types of radiation, chemotherapeutic drugs, and histone deacetylase inhibitors (Criswell *et al.*, 2005; Hoeller *et al.*, 2005; Liu *et al.*, 2009; Miyake *et al.*, 2000b; Trougakos *et al.*, 2004, 2005; Xie *et al.*, 2005; Zellweger *et al.*, 2003; Zhang *et al.*, 2005). sCLU function during cell exposure to genuine oxidants was also cytoprotective. Finally, in a series of studies in $CLU^{-/-}$ mice, CLU was found to confer protection from autoimmune myocardial damage (McLaughlin *et al.*, 2000), permanent focal cerebral ischemia (Wehrli *et al.*, 2001), heat-shock-mediated apoptosis in the testis (Bailey *et al.*, 2002), and progressive age-related glomerulopathy (Rosenberg *et al.*, 2002). In support to the mainstream notion that favors a prosurvival role for sCLU (see also chapter "Cell Protective Functions of

Secretory CLU (sCLU)"), specific inhibition of its biosynthesis by either antisense oligonucleotides (ASOs) or small-interfering RNA (siRNA) resulted in significant cell sensitization to apoptosis induced by oxidative stress (Viard et al., 1999), chemotherapeutic drugs (Hoeller et al., 2005; Miyake et al., 2000a; Trougakos et al., 2004), IR (Criswell et al., 2005), or TRAIL (Sallman et al., 2007). Finally, it was recently found in prostate cancer cells that sCLU promoted survival by activating the prosurvival phosphatidylinositol-3-kinase/Akt pathway (Ammar and Closset, 2008).

Interestingly, an increasing number of reports indicate that even in the absence of external stress effective sCLU knock down reduces cell proliferation and increases the rates of endogenous spontaneous apoptosis in several normal or cancer mammalian cell types (Kang et al., 2000; Moulson and Millis, 1999; Schlabach et al., 2008; Sensibar et al., 1995; Trougakos et al., 2004, 2009b; Zwain and Amato, 2000). These findings open new paths in sCLU research as they clearly demonstrate the fundamental role of sCLU in suppressing endogenous cellular apoptotic signals and in the maintenance of cell homeostasis.

Recent efforts have provided some mechanistic models and shed light to sCLU implication in the regulation of the apoptotic machinery. More specifically, it was shown that CLU inhibits apoptosis by specifically interacting with conformation-altered activated Bax in the mitochondria (Zhang et al., 2005). This study was corroborated by the recent finding that sCLU binds and thereby stabilizes the Ku70–Bax protein complex serving as a cytosol retention factor for Bax (Trougakos et al., 2009b). These facts along with sCLU localization, both extracellularly and in various intracellular compartments, indicate that apart from its induction by oxidants in order to chaperone damaged proteins and reduce cell damage sCLU may also exert its chaperone function (intracellularly or extracellularly) even in the absence of stress having a vital housekeeping role in the maintenance of cellular proteome stability. This assumption is in line with previous reports as, apart from Ku70 and Bax binding in normally growing cells (Trougakos et al., 2009b; Zhang et al., 2005), sCLU also binds the TGF-β type I/II receptors (Reddy et al., 1996), the cytosolic microtubule-destabilizing stathmin family protein SCLIP (Kang et al., 2005) and phosphorylated IκBα (Devauchelle et al., 2006).

B. Implication of Clusterin in States of Increased Organismal Oxidative Stress

During in vivo ageing, CLU expression levels were higher in the ventral and lateral prostate lobes of aged as compared to young rats (Lam et al., 2008; Lau et al., 2003; Omwancha et al., 2009). Moreover, aging in rats impaired the serum lipid profile and increased lipid peroxidation, PON1 activities, and the content of both PON1 and sCLU in HDL which suggests an HDL subfraction redistribution to protect low-density lipoproteins more effectively from oxidation (Thomàs-Moyà *et al.*, 2006). In humans, sCLU was upregulated from gestation to adults (Wong *et al.*, 1994) and accumulated in the serum of elder subjects (Trougakos *et al.*, 2002b). Also it was found to be induced during ageing of the pituitary gland (Ishikawa *et al.*, 2006), glial cells (Patel *et al.*, 2004), lymphocytes (Trougakos *et al.*, 2006b), and adenohypophysis (Doğan Ekici *et al.*, 2008). Considering that ageing is among the main risk factors for many pathologies including vascular diseases, neurodegeneration, chronic inflammation, and cancer it is not surprising that CLU has been implicated in all these pathological conditions.

During vascular damage, sCLU was found to accumulate in the human serum of diabetes type II patients or during myocardial infraction (Trougakos *et al.*, 2002b). Also, sCLU accumulated in the artery wall during atherosclerosis where it was proposed to protect against the oxidative stress associated with the development of the disease (Mackness *et al.*, 1997). sCLU prevented fetal cardiac myoblast apoptosis induced by ethanol (Li *et al.*, 2007) and protected cardiomyocytes against ischemic cell death via a complement-independent pathway (Krijnen *et al.*, 2005). It was recently proposed that during atherogenesis the protective role of CLU may be exerted by binding to enzymatically modified low-density lipoprotein to reduce fatty acid-mediated cytotoxicity (Schwarz *et al.*, 2008).

Normal adult brain is a major site of CLU mRNA synthesis in several mammalian species and a significant increase in either mRNA or CLU protein expression has been observed during neurodegenerative conditions as well as during injury or chronic inflammation of the brain (Calero *et al.*, 2005). sCLU was found to associate with aggregated amyloid beta-peptide in senile and diffuse plaques of AD (Kida *et al.*, 1995), while in cortex and hippocampus extracts CLU concentration was about 40% higher in AD than in control individuals (Oda *et al.*, 1994). The role of CLU during neurodegeneration appeared controversial. Specifically, sCLU preincubation with Aβ before adding the mixture to PC12 cells enhanced the oxidation properties of Aβ and promoted its cytotoxicity (Oda *et al.*, 1995). On the other hand, sCLU was shown to prevent *in vitro* the Aβ peptide aggregation, polymerization, amyloid fibril formation, and neurotoxicity (Boggs *et al.*, 1996; Matsubara, *et al.*, 1996) as well as the spontaneous aggregation of a synthetic peptide homologous to human prion protein (McHattie and Edington, 1999). These discrepancies may be partially explained by the recent observation that the proamyloidogenic effects of CLU relate to the CLU:client protein ratio as when the substrate protein is present at a very large molar excess CLU coincorporates with it into insoluble aggregates, whereas when CLU is present in higher levels it potently inhibits amyloid formation and provides substantial cytoprotection (Yerbury *et al.*, 2007) (see also chapter "The Chaperone Action of CLU and Its Putative Role in Quality Control of Extracellular Protein Folding"). Studies at *in vivo* models of PDAPP (homozygous for the APPV717F transgene)/CLU$^{-/-}$ mice showed a similar degree of Aβ deposition as compared with PDAPP/CLU$^{+/+}$ animals. However, the deposits were largely thioflavin negative in PDAPP/CLU$^{-/-}$ mice indicating that fewer amyloid deposits were formed (DeMattos *et al.*, 2002). Thus, according to this report, similarly to what previously proposed for Apolipoprotein E (Bales *et al.*, 1997), sCLU enhances Aβ toxicity by facilitating the conversion of soluble Aβ into amyloids. Intriguingly, it seems that Apolipoprotein E and CLU act cooperatively in the suppression of Aβ deposition as the double knockout PDAPP/ApoE$^{-/-}$/CLU$^{-/-}$ mice had an earlier onset and a significant increase in Aβ and amyloid deposition (DeMattos *et al.*, 2004).

Chronic inflammation results in increased oxidative stress and it thus represent an appropriate model to test CLU responsiveness to oxidant injury (see also chapter "CLU, Chronic Inflammation and Autoimmunity"). It was found that CLU expression levels correlated positively with the duration of juvenile dermatomyositis, a disease that introduces chronic inflammation in muscle (Chen *et al.*, 2008b), as well as that endotoxin, TNF-α, and IL-1 increased hepatic mRNA and serum protein levels of sCLU in Syrian hamsters (Hardardóttir *et al.*, 1994). Also CLU was upregulated following Venezuelan equine encephalitis virus infection (Sharma *et al.*, 2008) and found to exert a cytoprotective function in lung during leukocyte-induced injury (Heller *et al.*, 2003); in inflammatory exocrine pancreas (Savković *et al.*, 2007) and in glomerulonephritis (Rastaldi *et al.*, 2006). Finally, sCLU was induced during ventricular heart tissue damage but not in the hypertrophic tissue (Swertfeger *et al.*, 1996) while in acute sepsis and septic shock sCLU concentration in plasma was higher in septic patients as compared to controls (Kalenka, 2006). The cytoprotective role of sCLU is evident if we consider the observation that sCLU was highly upregulated in the serum of patients that survived sepsis and septic shock ($\sim$26.5- or 15-fold as compared to controls) whereas in nonsurvived patients sCLU upregulation was lower (5.9- to 3.1-fold) (Kalenka *et al.*, 2006).

In support to its proposed anti-inflammatory function, sCLU suppressed the pro-death activity of the inflammatory cytokine TNF-α in prostate cancer or mouse fibroblastic cells (Humphreys *et al.*, 1997; Sensibar *et al.*, 1995; Sintich *et al.*, 1999). Interestingly, in breast cancer cells treatment with TNF-α altered the biogenesis of endogenous CLU during apoptosis resulting in the appearance of a $\sim$50 kDa uncleaved, nonglycosylated, disulfide-linked protein isoform of CLU that accumulates in the nucleus (O'Sullivan *et al.*, 2003).

In certain experimental models basal CLU gene expression or induction following cell exposure to stress depended on NF-κB activity (Li *et al.*, 2002; Saura *et al.*, 2003) (see also chapter "Regulation of CLU Gene Expression by Oncogenes, Epigenetic Regulation, and the Role of CLU in Tumorigenesis"). On the other hand, CLU itself mostly exerts a negative effect on NF-κB pathway activity indicating the existence of a negative feedback loop. Specifically, during proteinuria CLU inhibited NF-κB-dependent Bcl-X$_L$ expression by stabilizating IkappaBalpha (Takase *et al.*, 2008). Similarly, in human neuroblastoma cells and murine embryonic fibroblasts CLU inhibited NF-κB by stabilizing IkappaBs (Santilli *et al.*, 2003), while in a model of mouse pancreatitis CLU reduced NF-κB activation and downstream expression of the NF-κB target genes TNF-α and MOB-1 (Savković *et al.*, 2007). In human synoviocytes CLU was found to suppress NF-κB-regulated cytokine production; it inhibited TNF-α-mediated NF-κB activation and interacted with phosphorylated IκB suppressing NF-κB nuclear translocation (Devauchelle *et al.*, 2006) (see also chapter "CLU, Chronic Inflammation and Autoimmunity").

Finally, CLU has been functionally implicated in all aspects of tumorigenesis including tumor formation and progression, metastasis, and chemoresistance acquisition (Shannan *et al.*, 2006; Trougakos and Gonos, 2002) (see also chapters "CLU and Prostate Cancer," "CLU and Breast Cancer," "CLU and Colon Cancer," and "CLU and Lung Cancer," this volume; and "Regulation of CLU Gene Expression by Oncogenes, Epigenetic Regulation, and the Role of CLU in Tumorigenesis," Vol. 105). Changes in CLU (mostly sCLU) expression have been documented in a broad variety of different human malignancies. According to studies in esophageal squamous cell carcinoma (Zhang *et al.*, 2003) and in low- and high-grade human prostate cancer (Scaltriti *et al.*, 2004a), tumor progression and malignancy correlate to decreased expression of CLU. Also high CLU expression levels were found to positively correlate with better survival in stage III serous ovarian adenocarcinomas (Partheen *et al.*, 2008) and cytoplasmic CLU expression is associated with longer survival in patients with non small cell lung cancer (Albert *et al.*, 2007). In other reported cases, however, CLU mRNA and protein upregulation was closely associated with disease progression. Specifically, higher CLU levels were detected in kidney (Parczyk *et al.*, 1994) and prostate carcinomas (Steinberg *et al.*, 1997), in anaplastic large-cell lymphomas (Wellmann *et al.*, 2000), in ovarian cancer (Hough *et al.*, 2001), in human hepatocellular carcinoma (Kang *et al.*, 2004), in primary melanoma and melanoma metastases (Hoeller *et al.*, 2005), in leukemic non-Hodgkin's lymphomas (Chow *et al.*, 2006), in endometrioid and cervical carcinoma (Abdul-Rahman *et al.*, 2007; Ahn *et al.*, 2008), in pancreatic endocrine tumors (Mourra *et al.*, 2007), and during seminal vesicle carcinogenesis progression (Ouyang *et al.*, 2001). sCLU has been characterized as a sensitive biomarker of murine and human intestinal tumors (Chen *et al.*, 2003), anaplastic large-cell lymphomas (Saffer *et al.*, 2002), colon (Chen *et al.*, 2008a; Pucci *et al.*, 2004, 2009) and breast carcinomas (Redondo *et al.*, 2002). Moreover, high CLU expression correlated to the aggressiveness of breast carcinomas (Redondo *et al.*, 2002) and with a poor outcome in stage II colorectal cancers (Kevans *et al.*, 2009). In human lung adenocarcinoma cell lines CLU positively regulated EMT and aggressive behavior through modulating ERK signaling and Slug expression (Chou *et al.*, 2009). In support, sCLU has been found to be a potentially new prognostic and predictive marker for tumor metastasis in hepatocellular (Lau *et al.*, 2006) and colon carcinomas (Pucci *et al.*, 2004, 2009). Protemic analysis in colorectal cancer patients revealed the presence of several CLU isoforms that were distinct in their isoelectric point and glycozylation pattern; some of these were induced in tumor samples and thus could have a potential value as colorectal tumor markers (Rodríguez-Piñeiro *et al.*, 2006). Finally, sCLU appeared to be associated with resistance to chemotherapy treatments *in vivo* (Zellweger *et al.*, 2003) and it is significantly overexpressed in multidrug resistant osteosarcoma cell lines (Lourda *et al.*, 2007) and in docetaxel-resistant prostate tumor cells (Patterson *et al.*, 2006) (see also chapter "CLU and Chemoresistance," Vol. 105). CLU was also found to interact with Paclitaxel and conferred Paclitaxel resistance in ovarian cancer (Park *et al.*, 2008). A direct proof of CLU involvement in tumor cells resistance to chemotherapy relates to the fact that direct inhibition of CLU biosynthesis by siRNA (Lourda *et al.*, 2007) or indirect by resveratrol (Patterson *et al.*, 2006; Sallman *et al.*, 2007) resensitized the drug-resistant cancer cells to the chemotherapeutic drugs used.

Intriguingly, although sCLU overexpression in Myc-transduced colonocytes decreased cell accumulation (Thomas Tikhonenko *et al.*, 2004), it promoted c-Myc-mediated transformation in Rat-1 cells *in vitro* and tumor progression *in vivo* into nude mice (Zhang *et al.*, 2005). In a recent study where the role of CLU during tumor development was investigated in mouse models of neuroblastoma, development of the disease and metastasis showed a negative correlation with CLU expression levels suggesting that, in this particular type of cancer, CLU is a tumor and metastasis suppressor gene (Chayka *et al.*, 2009). However, the picture gets more complicated if we also consider another recent study showing that loss of Nkx3.1, a transcriptional regulator and tumor suppressor gene in prostate cancer that is downregulated during early stages of prostate tumorigenesis, resulted in CLU aberrant overexpression in Nkx3.1-driven tumor initiation (Song *et al.*, 2009). This finding was investigated in independent loss-of-PTEN and c-myc overexpression prostate adenocarcinoma mouse models where Nkx3.1 loss is an early event or late event, respectively. CLU was found to be highly expressed in prostatic hyperplasia and intraepithelial neoplasia lesions that also lack

Nkx3.1 in the PTEN-deficient prostate, but not in similar lesions in the c-myc transgenic model. Meta-analysis of multiple prostate cancer gene expression data sets, including those from loss-of-Nkx3.1, loss-of-PTEN, c-myc overexpression, and constitutively active AKT prostate cancer models, further confirmed that genes associated with the loss-of-Nkx3.1 signature (like CLU) integrate with PTEN–AKT signaling pathways, but do not overlap with molecular changes associated with the c-myc signaling pathway. In human prostate tissue samples, loss of NKX3.1 expression and CLU overexpression are colocalized at sites of prostatic inflammatory atrophy, a possible very early stage of human prostate tumorigenesis (Song et al., 2009). Thus, in this model CLU seems to actively participate in the early phases of prostate cancer. Overall, the link between CLU expression and cancer progression seems to be complicated and may depend on the type of cancer, the oncogenic pathways activated, the endogenous CLU expression levels in the cell or animal model used, and on the tissue or cell-type studied.

C. What Can We Learn from the Functional Similarities Between Clusterin and Its Distal Homologues, the Small-Heat Shock Proteins?

HSPs induction upon exposure to environmental insults constitutes one of the most evolutionarily conserved stress responses known to the living world. Molecular chaperones are a large family of ubiquitous proteins that appeared early in evolution to form a defensive system in cells by sequestering damaged proteins and preventing their aggregation (Soti et al., 2005; Sun and McRae, 2005). HSPs comprise a group of related proteins classified into six major families according to their molecular weight (Martindale and Holbrook, 2002); from these, as mentioned, sCLU shares functional similarities with shSPs (Carver et al., 2003). Only few members of the shSPs family [e.g., the mammalian HSP27 (HSPB1) and aB-crystallin (HSPB5) proteins] display the features of a genuine HSP and are induced in response to stress (Arrigo et al., 2007).

As already described for CLU, the induction of HSPs in response to stress is largely mediated through transcriptional activation via HSF1 (Pirkkala et al., 2001). The common activating signal is likely to be oxidative stress and the consequent oxidative damage to proteins that alters their folding or introduce oxidative-inflammation conditions (Arrigo et al., 2007). In most mammals, shSPs are constitutively expressed in tissues with high rates of oxidative metabolism, including, the heart, type I and type IIa skeletal muscle fibers, brain, and oxidative regions of the kidney (Arrigo, 2001). The similar expression pattern of CLU in these tissues offers an additional line of defense in the organism to combat oxidative stress. However, the accumulation of a series of different regulatory elements in the CLU gene promoter makes CLU gene more sensitive than shSPs in the various environmental insults.

shSPs are molecular chaperones and function to protect proteome against denaturation or oxidative inactivation (Arrigo, 2001; Jolly and Morimoto, 2000). As in the case of CLU, insults that may activate shSPs include either direct oxidative damage (e.g., exposure to hydrogen peroxide or hypoxia reperfusion injury) or a variety of other stresses in which generation of oxygen-free radicals is indirect (e.g., chemotherapeutic drugs, heat stress, cytokines, etc.) (Graf and Jakob, 2002; Martindale and Holbrook, 2002). Elevated shSPs expression improves cell survival by reducing the oxidative damage to DNA, proteins, and lipids (Park *et al.*, 1998; Su *et al.*, 1999; Yamamoto *et al.*, 2000) and by decreasing the intracellular levels of free radicals and their derivatives as they modulate metabolism of glutathione to maintain it in a reduced state (Baek *et al.*, 2000; Preville *et al.*, 1999). Reportedly, shSPs preserve proteasome function (Ding and Keller, 2001), involve in cytoskeleton stabilization (Singh *et al.*, 2007), and suppress apoptotic signaling pathways downstream of mitochondria at the level of cytochrome *c* and apoptosome (Bruey *et al.*, 2000; Paul *et al.*, 2002). Also shSPs interact with caspases (Kamradt *et al.*, 2001) and suppress Bax, Bcl-X_S, and p53 polypeptides translocation to the mitochondria (Liu *et al.*, 2007; Mao *et al.*, 2004). As has been reported for sCLU, shSPs overexpression have been shown to enhance survival of cells and prevent apoptosis during a wide variety of stress conditions (Creagh *et al.*, 2000; Jolly and Morimoto, 2000). Apart from being expressed in stressed cells, some shSPs show a basal level of constitutive expression and act as in-house chaperone toward several fundamental cellular processes, such as cytoskeletal architecture, mutations masking, protein intracellular transport, intracellular redox homeostasis, and translation regulation (Arrigo *et al.*, 2007). The fact that sCLU shows a constitutive pattern of expression in many cell types indicates a similar profile of housekeeping functions as was recently proposed (Trougakos *et al.*, 2009b).

Additional functional similarities between sCLU and shSPs relate to their reported implication in various pathological conditions including *in vitro* senescence and ageing (Soti and Csermely, 2003), carcinogenesis, metastasis, and drug resistance (Aldrian *et al.*, 2002; Aoyama *et al.*, 1993; Chelouche-Lev *et al.*, 2004; Elpek *et al.*, 2003; Ungar *et al.*, 1994). Also, shSPs have been functionally implicated in heart attack and stroke (Hollander *et al.*, 2004; Ray *et al.*, 2001) where HSP27 transgenic mice are strongly protected against myocardial infarction and cerebral ischemia (Efthymiou *et al.*, 2004); in neurodegeneration (Renkawek *et al.*, 1994; Shinohara *et al.*, 1993; Yoo *et al.*, 2001); as well as in inflammatory diseases (Alford *et al.*, 2007).

Nonetheless, similarly to sCLU, sHSPs role in these diseases is dual as they may either exacerbate toxicity or function cytoprotectively (Calderwood and Ciocca, 2008).

A supposed main difference between sCLU and sHSPs seemed to be their differential localization. sHSPs are intracellular chaperones localized mainly in the cytosol and nucleus (Arrigo, 2007), whereas sCLU because of the ER-targeting hydrophobic signal peptide were expected to reside in the membranous compartments of the cell (e.g., ER, Golgi, secretory vesicles, and probably plasma membrane) as well as in the extracellular milieu (Calero et al., 2005; Shannan et al., 2006; Trougakos and Gonos, 2002). Most intriguingly, however, recent reports indicate that either under stress or even at normal conditions sCLU can escape, by a currently unknown mechanism, the membranous network of the cell and localize in the nucleus, cytosol, or mitochondria (Caccamo et al., 2006; Kang et al., 2005; O'Sullivan et al., 2003; Trougakos et al., 2009b; Zhang et al., 2005). Although further studies are needed to verify these findings, they raise the possibility that sCLU may represent the only chaperone that confers proteome stability both extra- and intracellularly.

VI. CONCLUDING REMARKS—PERSPECTIVES

From a first sight the function of CLU seems elusive the main cause being the opposing functions proposed in an array of various cell types and tissues. CLU seems to be an "*all in one*" super protein as its full-length version has signal peptides for both secretion and nuclear localization; it can bind a wide range of totally unrelated molecules and carries a wide array of different posttranslational modifications. On top of this, CLU gene promoter contains a collection of many distinct regulatory elements that have the potential to respond to even minute environmental changes or alterations in cellular homeostasis. Besides these unique properties, our protein homology searching in the existing databases revealed the existence of CLU-like protein sequences only in vertebrates. Also, a comparison of the CLU gene promoter DNA or the conventional full-length sCLU protein sequences among mammalian species showed a very high degree of conservation. These facts, along with the wide tissue distribution of CLU (Aronow et al., 1993), its implication in normal processes like development and differentiation (Garden et al., 1991; Itahana et al., 2007; O'Bryan et al., 1993; Trougakos et al., 2006a) and the absence of functional CLU gene polymorphisms in humans (Trougakos et al., 2006b; Tycko et al., 1996) clearly indicate that the protein has evolved in vertebrates to accomplish a function of fundamental biological importance.

After a period of debate the emerging consensus is that sCLU is mainly prosurvival, through its chaperone activity, whereas nCLU, when expressed, promotes cell death. Regarding sCLU, the idea that it cannot be "harmful *per se*" for the cells is mostly apparent by the facts that as mentioned it has a ubiquitous constitutive expression in most adult human tissues and it associates with cells surviving apoptosis during development and differentiation. The future discovery of more protein isoforms (e.g., those supposed to be primates specific; see also chapter "Regulation of CLU Gene Expression by Oncogenes, Epigenetic Regulation, and the Role of CLU in Tumorigenesis," Vol. 105) may add some complexity to that picture but we foresee that it will not dramatically alter our current ideas about CLU protein(s) sequence-derived function because all these isoforms are expected to differ in less than 10% of their N-terminus sequence. This difference although may impact on the localization pattern of the produced polypeptide(s) should not significantly affect the primary functional properties of the produced protein.

It is anticipated that the property of sCLU to chaperone unfolded damaged proteins most probably denotes its primary function. Based on the up to date observations and the CLU gene structure it is reasonable to argue that CLU is a sensitive biosensor of environmental insults and more particularly oxidative stress that is the driving force of most, if not all, age-related diseases (including cancer) where CLU has been functionally implicated. In these pathological conditions, the main mission of CLU would be to cease the deleterious effects of oxidative stress. If this hypothesis is true, then CLU gene expression and protein upregulation during ageing should reflect the general "oxidative status" of the subject rather than its chronological age. Indeed, although CLU gene expression levels in lymphocyte samples are higher in old donors as compared to their young counterparts, in similar samples from centenarians [a model of successful ageing characterized by low ROS load (Franceschi and Bonafe, 2003)], CLU gene expression levels were significantly lower than those found in the old donors (Trougakos *et al.*, 2006b).

On the basis of its chaperone activity and the supposed "beneficial" role of sCLU at the cellular level it is anticipated that there would be a selective (Darwinian) pressure for higher sCLU expression levels in both normal and pathological cells. However, the outcome at the organismal level may vary dramatically when we consider the added layers of complexity and the fact that what is "beneficial" at the cellular level may eventually become suicidal for the host; this argument mostly applies for cancer cells. Thus, sCLU implication in biological systems should be considered in a holistic view rather than focusing in a particular cellular state or disease (Fig. 2).

sCLU is neither a typical oncogene nor a typical tumor suppressor as according to current knowledge it cannot be classified as a classical

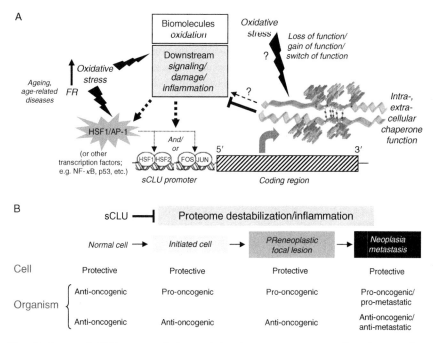

Fig. 2 Proposed CLU gene regulation and function as a response to altered oxygen-free radicals concentration. (A) As CLU promoter contains binding sites for HSF and AP-1 transcription factors complexes it is very sensitive to oxidative stress. CLU gene can also be regulated by the huge repertoire of molecules (e.g., cytokines, growth factors, etc.) produced as signaling (by)-products of oxidative stress and/or the accompanying inflammation. sCLU mostly suppress the toxic effects of oxidation ($\overline{\uparrow}$), although it cannot be excluded that in certain cellular contexts, it may have a protoxic effect ($\uparrow$). sCLU itself may also be subject to oxidative stress-induced modifications that may affect its function. (B) It is anticipated that at the cellular level sCLU mostly functions cytoprotectively. However, at the organismal level this property may suppress or promote tumor formation depending on various parameters such as the cell type or tissue characteristics; the endogenous levels of expression and isoforms (e.g., sCLU, nCLU) involved; the oncogenic pathways activated and the tissue microenvironment. FR, free radicals. (See Page 8 in Color Section at the back of the book.)

initiator or effector of cell cycle or death regulation (Trougakos *et al.*, 2009a). Its role seems to be more complicated as in early stages of cancer sCLU may indirectly act as a tumor suppressor due to its ability to suppress the accumulation of damaged proteins. However, during the initiation of neoplasia or at established tumors sCLU function may become oncogenic because of its ability to counteract apoptosis, either indirectly by actively involved in damage repair (Carver *et al.*, 2003) or directly by interfering with Bax pro-death activity (Trougakos *et al.*, 2009b; Zhang *et al.*, 2005), allowing thus cancer cells to acquire one of the main cancer hallmarks, namely resistance to apoptosis. Experimental evidence to support the "two-faced" character of CLU during tumorigenesis come from the observation that sCLU might act in a biphasic fashion during skin carcinogenesis as a tumor attenuator in early carcinogenesis and as an enhancer of the malignant phenotype in late carcinogenesis (Thomas-Tikhonenko *et al.*, 2004). Even in those cases where at the early phases of carcinogenesis the tumor microenvironment (e.g., aberrant oncogene activation) suppresses sCLU expression (Lund *et al.*, 2006) the selective pressure for CLU gene reactivation as the tumor evolves may eventually bypass the suppressing mechanisms allowing cells to reexpress sCLU. This theoretical model may explain recent reports showing that few nests of cancer cells, in an otherwise sCLU negative tumor, express high levels of the protein (Scaltriti *et al.*, 2004a) most probably due to *de novo* synthesis (Andersen *et al.*, 2007). It would be critical to address the question whether these tumor cells have a more aggressive phenotype as compared to the surrounding tissue. Apparently, the outcome of CLU contribution to tumor initiation and progression may also be affected by several other parameters including the cell type or tissue characteristics (i.e., the genetic background), the CLU endogenous expression levels and isoforms(s) implicated, the oncogenic stimuli and pathways activated, the tumor microenvironment and finally, the site of action.

In conclusion, the observations discussed herein, suggest that sCLU could be a valuable prognostic biomarker of increased organismal stress and disease progression. Also, sCLU, or other CLU protein isoforms, represent promising targets for the development of future therapeutic strategies against pathologies as diverse as cancers, inflammation, neurodegenerative diseases, or cataracts. Nevertheless, although it is anticipated that sCLU activation would be beneficial for the organism if we consider pathologies like atherosclerosis, degenerative diseases, or inflammation, the decision whether sCLU should be activated or inhibited in a certain tumor should take several parameters into consideration.

For instance, inhibition of NF-κB activation is now widely recognized as a valid drug-target strategy to combat inflammatory disease or cancer (Sarkar *et al.*, 2008). As CLU seems to be an NF-κB inhibitor its activation may has certain theoretical advantages in cancer therapy. sCLU activation may also provide therapeutic benefits in those tumors where sCLU exerts a tumor suppressing action (Chayka *et al.*, 2009). In addition, nCLU activation by natural compounds used either solely or in combination with chemotherapeutic drugs, may represent another therapeutic endeavor.

The alternative option of sCLU gene expression inhibition in tumor cells is currently exploited by OGX-011, a 2′-methoxyethyl-modified phosphorothioate ASO that is complementary to CLU mRNA (Chi *et al.*, 2008; Miyake *et al.*, 2000c). OGX-011 is in clinical trials for several types of cancer. Phase I trials showed that OGX-011 is well tolerated in patients, reduces CLU expression in prostate cancer, and enhances the apoptotic responses (Chi et al., 2005). According to a recent announcement of the final results from phase II clinical trials in patients receiving chemotherapy treatment for castrate-resistant prostate cancer (http://www.oncogenex. com/) treatment with an OGX-011/docetaxel combination was independently associated with improved overall patient survival in a preplanned multivariate analysis. Work in progress aims to finalize plans for two, phase III studies.

These latest exciting findings urge prompt investigation of various topics in CLU biology including: (a) its role in mitochondria and the additional activities (besides its chaperone function) needed to fight against apoptotic cell death; (b) its implication in the regulation of cell signaling pathways; (c) the molecular regulation, function, and localization of the various CLU protein isoforms in distinct cell types and tissues; (d) the isolation of novel extra- or intracellular binding partners; (e) its oxidation status during ageing or at the various age-related diseases and finally; (f) the detailed mapping and responsiveness to oxidative stress of the promoter(s) that regulate the various CLU isoforms. Last, but not least, the development of novel CLU overexpressing animal models along with the elucidation of CLU's crystal structure will provide precious information in our attempts to understand the function of this fascinating protein.

In any case, we should always bear in mind that in view of the integrative nature of living systems, the phenotypic outcome of stress-induced, homeostasis assuring proteins (like sCLU) in multifactorial diseases like cancer may be affected by several parameters resulting to either an antioncogenic or to a prooncogenic effect.

ACKNOWLEDGMENT

This work was supported by a "Kapodistrias" Grant of the University of Athens Special Account for Research Grants to IPT.

REFERENCES

Abdul-Rahman, P. S., Lim, B. K., and Hashim, O. H. (2007). Expression of high-abundance proteins in sera of patients with endometrial and cervical cancers: Analysis using 2-DE with silver staining and lectin detection 7methods. Electrophoresis 28, 1989–1996.

Ahn, H. J., Bae, J., Lee, S., Ko, J. E., Yoon, S., Kim, S. J., and Sakuragi, N. (2008). Differential expression of clusterin according to histological type of endometrial carcinoma. *Gynecol. Oncol.* **110,** 222–229.

Albert, J. M., Gonzalez, A., Massion, P. P., Chen, H., Olson, S. J., Shyr, Y., Diaz, R., Lambright, E. S., Sandler, A., Carbone, D. P., Putnam, J. B., Jr., Johnson, D. H., *et al.* (2007). Cytoplasmic clusterin expression is associated with longer survival in patients with resected non small cell lung cancer. *Cancer Epidemiol. Biomarkers Prev.* **16,** 1845–1851.

Aldrian, S., Trautinger, F., Frohlich, I., Berger, W., Micksche, M., and Kindas-Mugge, I. (2002). Overexpression of Hsp27 affects the metastatic phenotype of human melanoma cells *in vitro. Cell Stress Chaperones* **7,** 177–185.

Alexandrova, M., Bochev, P., Markova, V., Bechev, B., Popova, M., Danovska, M., and Simeonova, V. (2004). Dynamics of free radical processes in acute ischemic stroke: Influence on neurological status and outcome. *J. Clin. Neurosci.* **11,** 501–506.

Alexopoulos, E. C., Cominos, X., Trougakos, I. P., Lourda, M., Gonos, E. S., and Makropoulos, V. (2008). Biological monitoring of hexavalent chromium and serum levels of the senescence biomarker apolipoprotein j/clusterin in welders. *Bioinorg. Chem. Appl.* **2008,** 420578.

Alford, K. A., Glennie, S., Turrell, B. R., Rawlinson, L., Saklatvala, J., and Dean, J. L. (2007). Heat shock protein 27 functions in inflammatory gene expression and transforming growth factor-beta-activated kinase-1 (TAK1)-mediated signaling. *J. Biol. Chem.* **282,** 6232–6241.

Ammar, H., and Closset, J. L. (2008). Clusterin activates survival through the phosphatidylinositol 3-kinase/Akt pathway. *J. Biol. Chem.* **283,** 12851–12861.

Andersen, C. L., Schepeler, T., Thorsen, K., Birkenkamp-Demtröder, K., Mansilla, F., Aaltonen, L. A., Laurberg, S., and Ørntoft, T. F. (2007). Clusterin expression in normal mucosa and colorectal cancer. *Mol. Cell. Proteomics* **6,** 1039–1048.

Aoyama, A., Steiger, R. H., Frohli, E., Schafer, R., von Deimling, A., Wiestler, O. D., and Klemenz, R. (1993). Expression of aB-crystallin in human brain tumors. *Int. J. Cancer* **55,** 760–764.

Aronow, B. J., Lund, S. D., Brown, T. L., Harmony, J. A. K., and Witte, D. P. (1993). Apolipoprotein J expression at fluid-tissue interfaces: Potential role in barrier cytoprotection. *Proc. Natl. Acad. Sci. USA* **90,** 725–729.

Arrigo, A. P. (2001). Hsp27: Novel regulator of intracellular redox state. *IUBMB Life* **52,** 303–307.

Arrigo, A. P., Simon, S., Gibert, B., Kretz-Remy, C., Nivon, M., Czekalla, A., Guillet, D., Moulin, M., Diaz-Latoud, C., and Vicart, P. (2007). Hsp27 (HspB1) and alphaB-crystallin (HspB5) as therapeutic targets. *FEBS Lett.* **581,** 3665–3674.

Baek, S. H., Min, J. N., Park, E. M., Han, M. Y., Lee, Y. S., Lee, Y. J., and Park, Y. M. (2000). Role of small heat shock protein HSP25 in radioresistance and glutathione-redox cycle. *J. Cell. Physiol.* **183,** 100–107.

Bailey, R. W., Aronow, B., Harmony, J. A., and Griswold, M. D. (2002). Heat shock-initiated apoptosis is accelerated and removal of damaged cells is delayed in the testis of clusterin/ApoJ knock-out mice. *Biol. Reprod.* **66,** 1042–1053.

Balantinou, E., Trougakos, I. P., Chondrogianni, N., Margaritis, L. H., and Gonos, E. S. (2009). Transcriptional and post-translational regulation of clusterin by the two main cellular proteolytic pathways. *Free Radic. Biol. Med.* **46,** 1267–1274.

Bales, K. R., Verina, T., Dodel, R. C., Du, Y., Altstiel, L., Bender, M., Hyslop, P., Johnstone, E. M., Little, S. P., Cummins, D. J., Piccardo, P., Ghetti, B., *et al.* (1997). Lack of apolipoprotein E dramatically reduces amyloid beta-peptide deposition. *Nat. Genet.* **17,** 263–264.

Batchioglu, K., Mehmet, N., Ozturk, I. C., Yilmaz, M., Aydogdu, N., Erguvan, R., Uyumlu, B., Genc, M., and Karagozler, A. A. (2006). Lipid peroxidation and antioxidant status in stomach cancer. *Cancer Invest.* **24**, 18–21.

Bauerova, K., and Bezek, A. (1999). Role of reactive oxygen and nitrogen species in etiopathogenesis of rheumatoid arthritis. *Gen. Physiol. Biophys.* **18**, 15–20.

Beckman, K. B., and Ames, B. N. (1998). The free radical theory of aging matures. *Physiol. Rev.* **78**, 547–581.

Bellomo, G., and Mirabelli, F. (1992). Oxidative stress and cytoskeletal alterations. *Ann. N. Y. Acad. Sci.* **663**, 97–109.

Bettuzzi, S., Scorcioni, F., Astancolle, S., Davalli, P., Scaltriti, M., and Corti, A. (2002). Clusterin (SGP-2) transient overexpression decreases proliferation rate of SV40-immortalized human prostate epithelial cells by slowing down cell cycle progression. *Oncogene* **21**, 4328–4334.

Boggs, L. N., Fuson, K. S., Baez, M., Churgay, L., McClure, D., Becker, G., and May, P. C. (1996). Clusterin (Apo J) protects against *in vitro* amyloid-beta(1–40) neurotoxicity. *J. Neurochem.* **67**, 1324–1327.

Bruey, J. M., Ducasse, C., Bonniaud, P., Ravagnan, L., Susin, S. A., Diaz-Latoud, C., Gurbuxani, S., Arrigo, A. P., Kroemer, G., Solary, E., and Garrido, C. (2000). Hsp27 negatively regulates cell death by interacting with cytochrome *c*. *Nat. Cell Biol.* **2**, 645–652.

Caccamo, A. E., Desenzani, S., Belloni, L., Borghetti, A. F., and Bettuzzi, S. (2006). Nuclear clusterin accumulation during heat shock response: Implications for cell survival and thermotolerance induction in immortalized and prostate cancer cells. *J. Cell. Physiol.* **207**, 208–219.

Calderwood, S. K., and Ciocca, D. R. (2008). Heat shock proteins: Stress proteins with Janus-like properties in cancer. *Int. J. Hyperthermia* **24**, 31–39.

Calero, M., Rostagno, A., Frangione, B., and Ghiso, J. (2005). Clusterin and Alzheimer's disease. *Subcell. Biochem.* **38**, 273–298.

Campisi, J. (2005). Senescent cells, tumor suppression, and organismal aging: Good citizens, bad neighbors. *Cell* **120**, 513–522.

Carmeliet, P. (2000). Mechanisms of angiogenesis and arteriogenesis. *Nat. Med.* **6**, 389–395.

Carnevali, S., Luppi, F., D'Arca, D., Caporali, A., Ruggieri, M. P., Vettori, M. V., Caglieri, A., Astancolle, S., Panico, F., Davalli, P., Mutti, A., Fabbri, L. M., et al. (2006). Clusterin decreases oxidative stress in lung fibroblasts exposed to cigarette Smoke. *Am. J. Respir. Crit. Care Med.* **174**, 393–399.

Carrard, G., Bulteau, A. L., Petropoulos, I., and Friguet, B. (2002). Impairment of proteasome structure and function in aging. *Int. J. Biochem. Cell. Biol.* **34**, 1461–1474.

Carver, J. A., Rekas, A., Thorn, D. C., and Wilson, M. R. (2003). Small heat-shock proteins and clusterin: Intra- and extracellular molecular chaperones with a common mechanism of action and function? *IUBMB Life* **55**, 661–668.

Cavalli, L. R., and Liang, B. D. (1998). Mutagenesis, tumorigenicity, and apoptosis: Are the mitochondria involved? *Mutat. Res.* **398**, 19–26.

Chaiswing, L., Bourdeau-Heller, J. M., Zong, W., and Oberley, T. D. (2007). Characterization of redox state of two human prostate carcinoma cell lines with different degrees of aggressiveness. *Free Radic. Biol. Med.* **43**, 202–215.

Chayka, O., Corvetta, D., Dews, M., Caccamo, A. E., Piotrowska, I., Santilli, G., Gibson, S., Sebire, N. J., Himoudi, N., Hogarty, M. D., Anderson, J., Bettuzzi, S., et al. (2009). Clusterin, a haploinsufficient tumor suppressor gene in neuroblastomas. *J. Natl Cancer Inst.* **101**, 663–677.

Chelouche-Lev, D., Kluger, H. M., Berger, A. J., Rimm, D. L., and Price, J. E. (2004). aB-crystallin as a marker of lymph node involvement in breast carcinoma. *Cancer* **100**, 2543–2548.

Chen, F., Castranova, V., and Shi, X. (2001). New insights into the role of nuclear factor-B in cell growth regulation. *Am. J. Pathol.* **159**, 387–397.

Chen, X., Halberg, R. B., Ehrhardt, W. M., Torrealba, J., and Dover, W. F. (2003). Clusterin as a biomarker in murine and human intestinal neoplasia. *Proc. Natl. Acad. Sci. USA* **100**, 9530–9535.

Chen, T., Turner, J., McCarthy, S., Scaltriti, M., Bettuzzi, S., and Yeatman, T. J. (2004). Clusterin-mediated apoptosis is regulated by adenomatous polyposis coli and is p21 dependent but p53 independent. *Cancer Res.* **64**, 7412–7419.

Chen, X., Ehrhardt, W. M., Halberg, R. B., Aronow, B. J., and Dove, W. F. (2008a). Cellular expression patterns of genes upregulated in murine and human colonic neoplasms. *J. Histochem. Cytochem.* **56**, 433–441.

Chen, Y. W., Shi, R., Geraci, N., Shrestha, S., Gordish-Dressman, H., and Pachman, L. M. (2008b). Duration of chronic inflammation alters gene expression in muscle from untreated girls with juvenile dermatomyositis. *BMC Immunol.* **9**, 43.

Chen, J., Velalar, C. N., and Ruan, R. (2008c). Identifying the changes in gene profiles regulating the amelioration of age-related oxidative damages in kidney tissue of rats by the intervention of adult-onset calorie restriction. *Rejuvenation Res.* **11**, 757–763.

Chi, K. N., Eisenhauer, E., Fazli, L., Jones, E. C., Goldenberg, S. L., Powers, J., Tu, D., and Gleave, M. E. (2005). A phase I pharmacokinetic and pharmacodynamic study of OGX-011, a 2'-methoxyethyl antisense oligonucleotide to clusterin, in patients with localized prostate cancer. *J. Natl Cancer Inst.* **97**, 1287–1296.

Chi, K. N., Zoubeidi, A., and Gleave, M. E. (2008). Custirsen (OGX-011): A second-generation antisense inhibitor of clusterin for the treatment of cancer. *Expert Opin. Investig. Drugs* **17**, 1955–1962.

Chinenov, Y., and Kerppola, T. K. (2001). Close encounters of many kinds: Fos–Jun interactions that mediate transcription regulatory specificity. *Oncogene* **20**, 2438–2452.

Chondrogianni, N., and Gonos, E. S. (2005). Proteasome dysfunction in mammalian aging: Steps and factors involved. *Exp. Gerontol.* **40**, 931–938.

Chou, T. Y., Chen, W. C., Lee, A. C., Hung, S. M., Shih, N. Y., and Chen, M. Y. (2009). Clusterin silencing in human lung adenocarcinoma cells induces a mesenchymal-to-epithelial transition through modulating the ERK/Slug pathway. *Cell Signal.* **21**, 704–711.

Chow, K. U., Nowak, D., Kim, S. Z., Schneider, B., Komor, M., Boehrer, S., Mitrou, P. S., Hoelzer, D., Weidmann, E., and Hofmann, W. K. (2006). *In vivo* drugresponse in patients with leukemic non-Hodgkin's lymphomas is associated with *in vitro* chemosensitivity and gene expression profiling. *Pharmacol. Res.* **53**, 49–61.

Collado, M., and Serrano, M. (2006). Innovation: The power and the promise of oncogene-induced senescence markers. *Nat. Rev. Cancer* **6**, 472–476.

Creagh, E. M., Sheehan, D., and Cotter, T. G. (2000). Heat shock proteins—Modulators of apoptosis in tumour cells. *Leukemia* **14**, 1161–1173.

Criswell, T., Beman, M., Araki, S., Leskov, K., Cataldo, E., Mayo, L. D., and Boothman, D. A. (2005). Delayed activation of insulin-like growth factor-1 receptor/Src/MAPK/Egr-1 signaling regulates clusterin expression, a pro-survival factor. *J. Biol. Chem.* **280**, 14212–14221.

Debacq-Chainiaux, F., Borlon, C., Pascal, T., Royer, V., Eliaers, F., Ninane, N., Carrard, G., Friguet, B., de Longueville, F., Boffe, S., Remacle, J., and Toussaint, O. (2005). Repeated exposure of human skin fibroblasts to UVB at subcytotoxic level triggers premature senescence through the TGF-beta1 signaling pathway. *J. Cell Sci.* **118**, 743–758.

Debure, L., Vayssiere, J.L, Rincheval, V., Loison, F., Le Drean, Y., and Michel, D. (2003). Intracellular clusterin causes juxtanuclear aggregate formation and mitochondrial alteration. *J. Cell Sci.* **116**, 3109–3121.

DeMattos, R. B., O'dell, M. A., Parsadanian, M., Taylor, J. W., Harmony, J. A. K., Bales, K. R., Paul, S. M., Aronow, B. J., and Holtzman, D. M. (2002). Clusterin promotes amyloid plaque formation and is critical for neuritic toxicity in a mouse model of Alzheimer's disease. *Proc. Natl. Acad. Sci. USA* **99**, 10843–10848.

DeMattos, R. B., Cirrito, J. R., Parsadanian, M., May, P. C., O'Dell, M. A., Taylor, J. W., Harmony, J. A., Aronowv, B. J., Bales, K. R., Paul, S. M., and Holtzman, D. M. (2004). ApoE and clusterin cooperatively suppress Abeta levels and deposition. Evidence that ApoE regulates extracellular Abeta metabolism *In Vivo. Neuron* **41**, 193–202.

Demple, B., and Harrison, L. (1994). Repair of oxidative damage to DNA. *Annu. Rev. Biochem.* **63**, 915–948.

de Silva, H. V., Stuart, W. D., Duvic, C. R., Wetterau, J. R., Ray, M. J., Ferguson, D. G., Albers, H. W., Smith, W. R., and Harmony, J. A. (1990). A 70-kDa apolipoprotein designated ApoJ is a marker for subclasses of human plasma high density lipoproteins. *J. Biol. Chem.* **265**, 13240–13247.

Devauchelle, V., Essabani, A., De Pinieux, G., Germain, S., Tourneur, L., Mistou, S., Margottin-Goguet, F., Anract, P., Migaud, H., Le Nen, D., Lequerré, T., Saraux, A., *et al.* (2006). Characterization and functional consequences of underexpression of clusterin in rheumatoid arthritis. *J. Immunol.* **177**, 6471–6479.

Devi, G. S., Prasad, M. H., Saraswathi, I., Raghu, D., Rao, D. N., and Reddy, P. P. (2000). Free radicals antioxidant enzymes and lipid peroxidation in different types of leukemia. *Clin. Chim. Acta* **293**, 53–62.

de Visser, K. E., Korets, L. V., and Coussens, L. M. (2005). *De novo* carcinogenesis promoted by chronic inflammation is B lymphocyte dependent. *Cancer Cell* **7**, 411–423.

Ding, Q., and Keller, J. N. (2001). Proteasome inhibition in oxidative stress neurotoxicity: Implications for heat shock proteins. *J. Neurochem.* **77**, 1010–1017.

Doğan Ekici, A. I., Eren, B., Türkmen, N., Comunoğlu, N., and Fedakar, R. (2008). Clusterin expression in non-neoplastic adenohypophyses and pituitary adenomas: Cytoplasmic clusterin localization in adenohypophysis is related to aging. *Endocr. Pathol.* **19**, 47–53.

Dreher, D., and Junod, A. F. (1996). Role of oxygen free radicals in cancer development. *Eur. J. Cancer* **32A**, 30–38.

Dröge, W. (2002). Free radicals in the physiological control of cell function. *Physiol. Rev.* **82**, 47–95.

Dröge, W. (2003). Oxidative stress and aging. *Adv. Exp. Med. Biol.* **543**, 191–200.

Dröge, W., and Schipper, H. M. (2007). Oxidative stress and aberrant signaling in aging and cognitive decline. *Aging Cell* **6**, 361–370.

Dumont, P., Burton, M., Chen, Q. M., Gonos, E. S., Frippiat, C., Mazarati, J. B., Eliaers, F., Remacle, J., and Toussaint, O. (2000). Induction of replicative senescence biomarkers by sublethal oxidative stresses in normal human fibroblast. *Free Radic. Biol. Med.* **28**, 361–373.

Dumont, P., Chainiaux, F., Eliaers, F., Petropoulou, C., Koch-Brandt, C., Gonos, E. S., and Toussaint, O. (2002). Overexpression of Apolipoprotein J in human fibroblasts protects against cytotoxicity and premature senescence induce by ethanol and tert-butylhydroperoxide. *Cell Stress Chaperones* **7**, 23–35.

Dyrks, T., Dyrks, E., Masters, C. L., and Beyreuther, K. (1993). Amyloidogenicity of rodent and human beta A4 sequences. *FEBS Lett.* **324**, 231–236.

Efthymiou, C. A., Mocanu, M. M., de Belleroche, J., Wells, D. J., Latchmann, D. S., and Yellon, D. M. (2004). Heat shock protein 27 protects the heart against myocardial infarction. *Basic Res. Cardiol.* **99**, 392–394.

Elchuri, S., Oberley, T. D., Qi, W., Eisenstein, R. S., Jackson Roberts, L., Van Remmen, H., Epstein, C. J., and Huang, T. T. (2005). CuZnSOD deficiency leads to persistent and widespread oxidative damage and hepatocarcinogenesis later in life. *Oncogene* **24**, 367–380.

Elpek, G. O., Karaveli, S., Simsek, T., Keles, N., and Aksoy, N. H. (2003). Expression of heat-shock proteins hsp27, hsp70 and hsp90 in malignant epithelial tumour of the ovaries. *Apmis* **111**, 523–530.

Farrow, B., and Evers, B. M. (2002). Inflammation and the development of pancreatic cancer. *Surg. Oncol.* **10**, 153–169.

Federico, A., Morgillo, F., Tuccillo, C., Ciardiello, F., and Loguercio, C. (2007). Chronic inflammation and oxidative stress in human carcinogenesis. *Int. J. Cancer* **121**, 2381–2386.

Finkel, T., and Holbrook, N. J. (2000). Oxidants, oxidative stress and the biology of ageing. *Nature* **408**, 239–247.

Floyd, R. A., West, M. S., Eneff, K. L., Schneider, J. E., Wong, P. K., Tingey, D. T., and Hogsett, W. E. (1990). Conditions influencing yield and analysis of 8-hydroxy-20-deoxyguanosine in oxidatively damaged DNA. *Anal. Biochem.* **188**, 155–158.

Fraga, C. G., Shigenaga, M. K., Park, J. W., Degan, P., and Ames, B. N. (1990). Oxidative damage to DNA during aging: 8-hydroxy-20-deoxyguanosine in rat organ DNA and urine. *Proc. Natl. Acad. Sci. USA* **87**, 4533–4537.

Franceschi, C., and Bonafe, M. (2003). Centenarians as a model for healthy aging. *Biochem. Soc. Trans.* **31**, 457–461.

Francia, P., delli Gatti, C., Bachschmid, M., Martin-Padura, I., Savoia, C., Migliaccio, E., Pelicci, P. G., Schiavoni, M., Luscher, T. F., Volpe, M., and Cosentino, F. (2004). Deletion of p66shc gene protects against age-related endothelial dysfunction. *Circulation* **110**, 2889–2895.

Frippiat, C., Chen, Q. M., Zdanov, S., Magalhaes, J. P., Remacle, J., and Toussaint, O. (2001). Subcytotoxic H2O2 stress triggers a release of transforming growth factor-β1 which induces biomarkers of cellular senescence in human diploid fibroblasts. *J. Biol. Chem.* **276**, 2531–2537.

Garden, G. A., Bothwell, M., and Rubel, E. W. (1991). Lack of correspondence between mRNA expression for a putative cell death molecule (SGP-2) and neuronal cell death in the central nervous system. *J. Neurobiol.* **22**, 590–604.

Golden, T. R., Hinerfeld, D. A., and Melov, S. (2002). Oxidative stress and aging: Beyond correlation. *Aging Cell* **2**, 117–123.

Graf, P. C., and Jakob, U. (2002). Redox-regulated molecular chaperones. *Cell. Mol. Life Sci.* **59**, 1624–16231.

Grune, T., Jung, T., Merker, K., and Davies, K. J. (2004). Decreased proteolysis caused by protein aggregates, inclusion bodies, plaques, lipofuscin, ceroid, and "aggresomes" during oxidative stress, aging, and disease. *Int. J. Biochem. Cell Biol.* **36**, 2519–2530.

Gunter, M. J., and Leitzmann, M. F. (2006). Obesity and colorectal cancer: Epidemiology, mechanisms and candidate genes. *J. Nutr. Biochem.* **17**, 145–156.

Gupte, A., and Mumper, R. J. (2009). Elevated copper and oxidative stress in cancer cells as a target for cancer treatment. *Cancer Treat. Rev.* **35**, 32–46.

Halazonetis, T. D., Gorgoulis, V. G., and Bartek, J. (2008). An oncogene-induced DNA damage model for cancer development. *Science* **319**, 1352–1355.

Halliwell, B. (2007). Oxidative stress and cancer: Have we moved forward? *Biochem. J.* **401**, 1–11.

Han, B. H., DeMattos, R. B., Dugan, L. L., Kim-Han, J. S., Brendza, R. P., Fryer, J. D., Kierson, M., Cirrito, J., Quick, K., Harmony, J. A., Aronow, B. J., and Holtzman, D. M. (2001). Clusterin contributes to caspase-3-independent brain injury following neonatal hypoxia-ischemia. *Nat. Med.* **7**, 338–743.

Hanahan, D., and Weinberg, R. A. (2000). The hallmarks of cancer. *Cell* **100**, 57–70.

Hardardóttir, I., Kunitake, S. T., Moser, A. H., Doerrler, W. T., Rapp, J. H., Grünfeld, C., and Feingold, K. R. (1994). Endotoxin and cytokines increase hepatic messenger RNA levels and serum concentrations of apolipoprotein J (clusterin) in Syrian hamsters. *J. Clin. Invest.* **94**, 1304–1309.

Hekimi, S., and Guarente, L. (2003). Genetics and the specificity of the aging process. *Science* **299**, 1351–1354.

Heller, A. R., Fiedler, F., Braun, P., Stehr, S. N., Bödeker, H., and Koch, T. (2003). Clusterin protects the lung from leukocyte-induced injury. *Shock* **20**, 166–170.

Hill, M. F., and Singal, P. K. (1997). Right and left myocardial antioxidant responses during heart failure subsequent to myocardial infarction. *Circulation* **96**, 2414–2420.

Ho, E., and Bray, T. M. (1999). Antioxidants,NFkappaB activation, and diabetogenesis. *Proc. Soc. Exp. Biol. Med.* **222**, 205–213.

Hoeller, C., Pratscher, B., Thallinger, C., Winter, D., Fink, D., Kovacic, B., Sexl, V., Wacheck, V., Gleave, M. E., Pehamberger, H., and Jansen, B. (2005). Clusterin regulates drug-resistance in melanoma cells. *J. Invest. Dermatol.* **124**, 1300–1307.

Hollander, J. M., Martin, J. L., Belke, D. D., Scott, B. T., Swanson, E., Krishnamoorthy, V., and Dillmann, W. H. (2004). Overexpression of wildtype heat shock protein 27 and a nonphosphorylatable heat shock protein 27 mutant protects against ischemia/reperfusion injury in a transgenic mouse model. *Circulation* **110**, 3544–3552.

Hough, C. D., Cho, K. R., Zonderman, A. B., Schwartz, D. R., and Morin, P. J. (2001). Coordinately up-regulated genes in ovarian cancer. *Cancer Res.* **61**, 3869–3876.

Humphreys, D., Hochgrebe, T. T., Easterbrook-Smith, S. B., Tenniswood, M. P., and Wilson, M. R. (1997). Effects of clusterin overexpression on TNFalpha- and TGFbeta-mediated death of L929 cells. *Biochemistry* **36**, 15233–15243.

Humphreys, D. T., Carver, J. A., Easterbrook-Smith, S. B., and Wilson, M. R. (1999). Clusterin has chaperone-like activity similar to that of small heat shock proteins. *J. Biol. Chem.* **274**, 6875–6881.

Ishikawa, T., Zhu, B. L., Li, D. R., Zhao, D., Michiue, T., and Maeda, H. (2006). Age-dependent increase of clusterin in the human pituitary gland. *Leg. Med. (Tokyo)* **8**, 28–33.

Itahana, Y., Piens, M., Sumida, T., Fong, S., Muschler, J., and Desprez, P. Y. (2007). Regulation of clusterin expression in mammary epithelial cells. *Exp. Cell Res.* **313**, 943–951.

Janero, D. R. (1990). Malondialdehyde and thiobarbituric acid reactivity as diagnostic indices of lipid peroxidation and peroxidative tissue injury. *Free Radic. Biol. Med.* **9**, 515–540.

Janig, E., Haslbeck, M., Aigelsreiter, A., Braun, N., Unterthor, D., Wolf, P., Khaskhely, N. M., Buchner, J., Denk, H., and Zatloukal, K. (2007). Clusterin associates with altered elastic fibers in human photoaged skin and prevents elastin from ultraviolet-induced aggregation *in vitro*. *Am. J. Pathol.* **171**, 1474–1482.

Jenne, D. E., and Tschopp, J. (1989). Molecular structure and functional characterization of a human complement cytolysis found in blood and seminal plasma: Identity to sulfated glycoprotein 2, a constituent of rat testis fluid. *Proc. Natl. Acad. Sci. USA* **86**, 7123–7127.

Jenner, P. (2003). Oxidative stress in Parkinson's disease. *Ann. Neurol.* **53**, S26–S36.

Jin, G., and Howe, P. H. (1997). Regulation of clusterin gene expression by transforming growth factor α. *J. Biol. Chem.* **272**, 26620–26626.

Jo, H., Jia, Y., Subramanian, K. K., Hattori, H., and Luo, H. R. (2008). Cancer cell-derived clusterin modulates the phosphatidylinositol 3′-kinase-Akt pathway through attenuation of insulin-like growth factor 1 during serum deprivation. *Mol. Cell. Biol.* **28**, 4285–4299.

Jolly, C., and Morimoto, R. I. (2000). Role of the heat shock response and molecular chaperones in oncogenesis and cell death. *J. Natl Cancer Inst.* **92**, 1564–1572.

Kalenka, A., Feldmann, R. E., Jr., Otero, K., Maurer, M. H., Waschke, K. F., and Fiedler, F. (2006). Changes in the serum proteome of patients with sepsis and septic shock. *Anesth. Analg.* **103**, 1522–1526.

Kalka, K., Ahmad, N., Criswell, T., Boothman, D., and Mukhtar, H. (2000). Up-regulation of clusterin during phthalocyanine 4 photodynamic therapy-mediated apoptosis of tumor cells and ablation of mouse skin tumors. *Cancer Res.* **60**, 5984–5987.

Kamradt, M. C., Chen, F., and Cryns, V. L. (2001). The small heat shock protein alpha B-crystallin negatively regulates cytochrome *c*- and caspase-8-dependent activation of caspase-3 by inhibiting its autoproteolytic maturation. *J. Biol. Chem.* **276**, 16059–16063.

Kang, S. W., Lim, S. W., Choi, S. H., Shin, K. H., Chun, B. G., Park, I. S., and Min, B. H. (2000). Antisense oligonucleotide of clusterin mRNA induces apoptotic cell death and prevents adhesion of rat ASC-17D Sertoli cells. *Mol. Cells* **10**, 193–198.

Kang, Y. K., Hong, S. W., Lee, H., and Kim, W. H. (2004). Overexpression of clusterin in human hepatocellular carcinoma. *Hum. Pathol.* **35,** 1340–1346.

Kang, S. W., Shin, Y. J., Shim, Y. J., Jeong, S. Y., Park, I. S., and Min, B. H. (2005). Clusterin interacts with SCLIP (SCG10-like protein) and promotes neurite outgrowth of PC12 cells. *Exp. Cell Res.* **309,** 305–315.

Katsiki, M., Trougakos, I. P., Chondrogianni, N., Alexopoulos, E. C., Makropoulos, V., and Gonos, E. S. (2004). Alterations of senescence biomarkers in human cells by exposure to CrVI *in vivo* and *in vitro*. *Exp. Gerontol.* **39,** 1079–1087.

Kevans, D., Foley, J., Tenniswood, M., Sheahan, K., Hyland, J., O'Donoghue, D., Mulcahy, H., and O'Sullivan, J. (2009). High clusterin expression correlates with a poor outcome in stage II colorectal cancers. *Cancer Epidemiol. Biomarkers Prev.* **18,** 393–399.

Kida, E., Choi-Miura, N. H., and Wisniewski, K. E. (1995). Deposition of apolipoproteins E and J in senile plaques is topographically determined in both Alzheimer's disease and Down's syndrome brain. *Brain Res.* **685,** 211–216.

Kim, J. H., Kim, J. H., Yu, Y. S., Min, B. H., and Kim, K. W. (2007a). The role of clusterin in retinal development and free radical damage. *Br. J. Ophthalmol.* **91,** 1541–1546.

Kim, J. H., Yu, Y. S., Kim, J. H., Kim, K. W., and Min, B. H. (2007b). The role of clusterin in *in vitro* ischemia of human retinal endothelial cells. *Curr. Eye Res.* **32,** 693–698.

Kirkwood, T. B. (2002). Evolution of ageing. *Mech. Ageing Dev.* **123,** 737–745.

Klatt, P., Molina, E. P., DeLacoba, M. G., Padilla, C. A., Martinez-Galesto, E., Barcena, J. A., and Lamas, S. (1999). Redox regulation of c-Jun binding by reversible glutathiolation. *Faseb J.* **13,** 1481–1490.

Klaunig, J. E., and Kamendulis, L. M. (2004). The role of oxidative stress in carcinogenesis. *Annu. Rev. Pharmacol. Toxicol.* **44,** 239–267.

Klokov, D., Criswell, T., Leskov, K. S., Araki, S., Mayo, L., and Boothman, D. A. (2004). IR-inducible clusterin gene expression: A protein with potential roles in ionizing radiation-induced adaptive responses, genomic instability, and bystander effects. *Mutat. Res.* **568,** 97–110.

Kondo, S., Toyokuni, S., Iwasa, Y., Tanaka, T., Onodera, H., Hiai, H., and Imamura, M. (1999). Persistent oxidative stress in human colorectal carcinoma, but not adenoma. *Free Radic. Biol. Med.* **27,** 401–410.

Krijnen, P. A., Cillessen, S. A., Manoe, R., Muller, A., Visser, C. A., Meijer, C. J., Musters, R. J., Hack, C. E., Aarden, L. A., and Niessen, H. W. (2005). Clusterin: A protective mediator for ischemic cardiomyocytes? *Am. J. Physiol. Heart Circ. Physiol.* **289,** H2193–H2202.

Lam, Y. W., Tam, N. N., Evans, J. E., Green, K. M., Zhang, X., and Ho, S. M. (2008). Differential proteomics in the aging Noble rat ventral prostate. *Proteomics* **8,** 2750–2763.

Lane, D. P. (1992). Cancer. p53, guardian of the genome. *Nature* **358,** 15–16.

Lane, N. (2003). A unifying view of ageing and disease: The doubleagent theory. *J. Theor. Biol.* **225,** 531–540.

Lau, K. M., Tam, N. N., Thompson, C., Cheng, R. Y., Leung, Y. K., and Ho, S. M. (2003). Age-associated changes in histology and gene-expression profile in the rat ventral prostate. *Lab. Invest.* **83,** 743–757.

Lau, S. H., Sham, J. S., Xie, D., Tzang, C. H., Tang, D., Ma, N., Hu, L., Wang, Y., Wen, J. M., Xiao, G., Zhang, W. M., Lau, G. K., et al. (2006). Clusterin plays an important role in hepatocellular carcinoma metastasis. *Oncogene* **25,** 1242–1250.

Leskov, K. S., Klokov, D. Y., Li, J., Kinsella, T. J., and Boothman, D. A. (2003). Synthesis and functional analysis of nuclear clusterin: A cell death protein. *J. Biol. Chem.* **278,** 11590–11600.

Li, R. K., Sole, M. J., Mickle, D. A., Schimmer, J., and Goldstein, D. (1998). Vitamin E and oxidative stress in the heart of the cardiomyopathic syrian hamster. *Free Radic. Biol. Med.* **24,** 252–258.

Li, X., Massa, P. E., Hanidu, A., Peet, G. W., Aro, P., Savitt, A., Mische, S., Li, J., and Marcu, K. B. (2002). IKKalpha, IKKbeta, and NEMO/IKKgamma are each required for the NF-kappa B-mediated inflammatory response program. *J. Biol. Chem.* **277**, 45129–45140.

Li, Y., Sagar, M. B., Wassler, M., Shelat, H., and Geng, Y. J. (2007). Apolipoprotein-J prevention of fetal cardiac myoblast apoptosis induced by ethanol. *Biochem. Biophys. Res. Commun.* **357**, 157–161.

Liu, S., Li, J., Tao, Y., and Xiao, X. (2007). Small heat shock protein alphaB-crystallin binds to p53 to sequester its translocation to mitochondria during hydrogen peroxide-induced apoptosis. *Biochem. Biophys. Res. Commun.* **354**, 109–114.

Liu, T., Liu, P. Y., Tee, A. E., Haber, M., Norris, M. D., Gleave, M. E., and Marshall, G. M. (2009). Over-expression of clusterin is a resistance factor to the anti-cancer effect of histone deacetylase inhibitors. *Eur. J. Cancer* **45**, 1846–1854.

Loft, S., and Poulsen, H. E. (1996). Cancer risk and oxidative DNA damage in man. *J. Mol. Med. JMM* **74**, 297–312.

Loison, F., Debure, L., Nizard, P., le Goff, P., Michel, D., and le Drean, Y. (2006). Up-regulation of the clusterin gene after proteotoxic stress: Implication of HSF1-HSF2 heterocomplexes. *Biochem. J.* **395**, 223–231.

Lourda, M., Trougakos, I. P., and Gonos, E. S. (2007). Development of resistance to chemotherapeutic drugs in human osteosarcoma cell lines largely depends on up-regulation of clusterin/apolipoprotein J. *Int. J. Cancer* **120**, 611–622.

Lund, P., Weisshaupt, K., Mikeska, T., Jammas, D., Chen, X., Kuban, R. J., Ungethüm, U., Krapfenbauer, U., Herzel, H. P., Schäfer, R., Walter, J., and Sers, C. (2006). Oncogenic HRAS suppresses clusterin expression through promoter hypermethylation. *Oncogene* **25**, 4890–4903.

Mackness, B., Hunt, R., Durrington, P. N., and Mackness, M. I. (1997). Increased immunolocalization of paraoxonase, clusterin, and apolipoprotein A-I in the human artery wall with the progression of atherosclerosis. *Arterioscler. Thromb. Vasc. Biol.* **17**, 1233–1338.

Mancuso, C., Bates, T. E., Butterfield, D. A., Calafato, S., Cornelius, C., De Lorenzo, A., Dinkova Kostova, A. T., and Calabrese, V. (2007). Natural antioxidants in Alzheimer's disease. *Expert Opin. Investig. Drugs* **16**, 1921–1931.

Mantovani, A. (2009). Cancer: Inflaming metastasis. *Nature* **457**, 36–37.

Mantovani, A., Allavena, P., Sica, A., and Balkwill, F. (2008). Cancer-related inflammation. *Nature* **454**, 436–444.

Mao, Y. W., Liu, J. P., Xiang, H., and Li, D. W. (2004). Human alphaA- and alphaB-crystallins bind to Bax and Bcl-X(S) to sequester their translocation during staurosporine-induced apoptosis. *Cell Death Differ.* **11**, 512–526.

Mariani, E., Polidori, M. C., Cherubini, A., and Mecocci, P. (2005). Oxidative stress in brain aging, neurodegenerative and vascular diseases: An overview. *J. Chromatogr. B Analyt. Technol. Biomed. Life Sci.* **827**, 65–75.

Martindale, J. L., and Holbrook, N. J. (2002). Cellular response to oxidative stress: Signaling for suicide and survival. *J. Cell Physiol.* **192**, 1–15.

Mates, J. M., Perez-Gomez, C., and De Castro, I. N. (1999). Antioxidant enzymes and human diseases. *Clin. Biochem.* **32**, 595–603.

Matsubara, E., Soto, C., Governale, S., Frangione, B., and Ghiso, J. (1996). Apolipoprotein J and Alzheimer's amyloid beta solubility. *Biochem. J.* **316**, 671–679.

McCall, M. R., and Frei, B. (1999). Can antioxidant vitamins materially reduce oxidative damage in humans? *Free Radic. Biol. Med.* **26**, 1034–1053.

McHattie, S., and Edington, N. (1999). Clusterin prevents aggregation of neuropeptide 106–126 *in Vitro*. *Biochem. Biophys. Res. Commun.* **259**, 336–340.

McLaughlin, L., Zhu, G., Mistry, M., Ley-Ebert, C., Stuart, W. D., Florio, C. J., Groen, P. A., Witt, S. A., Kimball, T. R., Witte, D. P., Harmony, J. A., and Aronow, B. J. (2000). Apolipoprotein J/clusterin limits the severity of murine autoimmune myocarditis. *J. Clin. Invest.* **106**, 1105–1113.

Michel, D., Chatelain, G., Herault, Y., and Brun, G. (1995). The expression of the avian clusterin gene can be driven by two alternative promoters with distinct regulatory elements. *Eur. J. Biochem.* **229,** 215–223.

Migliaccio, E., Giorgio, M., Mele, S., Pelicci, G., Reboldi, P., Pandolfi, P. P., Lanfrancone, L., and Pelicci, P. G. (1999). The p66shc adaptor protein controls oxidative stress response and life span in mammals. *Nature* **402,** 309–313.

Minelli, A., Bellezza, I., Conte, C., and Culig, Z. (2009). Oxidative stress-related aging: A role for prostate cancer? *Biochim. Biophys. Acta* **1795,** 83–91.

Miyake, H., Chi, K. N., and Gleave, M. E. (2000a). Antisense TRPM-2 oligodeoxynucleotides chemosensitize human androgen-independent PC-3 prostate cancer cells both *in vitro* and *in vivo*. *Clin. Cancer Res.* **6,** 1655–1663.

Miyake, H., Nelson, C., Rennie, P. S., and Gleave, M. E. (2000b). Acquisition of chemoresistant phenotype by overexpression of the antiapoptotic gene testosterone-repressed prostate message-2 in prostate cancer xenograft models. *Cancer Res.* **60,** 2547–2554.

Miyake, H., Nelson, C., Rennie, P. S., and Gleave, M. E. (2000c). Testosterone-repressed prostate message-2 is an antiapoptotic gene involved in progression to androgen independence in prostate cancer. *Cancer Res.* **60,** 170–176.

Miyake, H., Hara, I., Gleave, M. E., and Eto, H. (2004). Protection of androgen-dependent human prostate cancer cells from oxidative stress-induced DNA damage by overexpression of clusterin and its modulation by androgen. *Prostate* **61,** 318–323.

Mollah, Z. U., Pai, S., Moore, C., O'Sullivan, B. J., Harrison, M. J., Peng, J., Phillips, K., Prins, J. B., Cardinal, J., and Thomas, R. (2008). Abnormal NF-{kappa}B function characterizes human type 1 diabetes dendritic cells and monocytes. *J. Immunol.* **180,** 3166–3175.

Moretti, R. M., Marelli, M. M., Mai, S., Cariboni, A., Scaltriti, M., Bettuzzi, S., and Limonta, P. (2007). Clusterin isoforms differentially affect growth and motility of prostate cells: Possible implications in prostate tumorigenesis. *Cancer Res.* **67,** 10325–10333.

Morimoto, R. I. (2002). Dynamic remodeling of transcription complexes by molecular chaperones. *Cell* **110,** 281–284.

Moulson, C. L., and Millis, A. J. (1999). Clusterin (Apo J) regulates vascular smooth muscle cell differentiation *in vitro*. *J. Cell. Physiol.* **180,** 355–364.

Mourra, N., Couvelard, A., Tiret, E., Olschwang, S., and Flejou, J. F. (2007). Clusterin is highly expressed in pancreatic endocrine tumours but not in solid pseudopapillary tumours. *Histopathology* **50,** 331–337.

Nizard, P., Tetley, S., Le Dréan, Y., Watrin, T., Le Goff, P., Wilson, M. R., and Michel, D. (2007). Stress-induced retrotranslocation of clusterin/ApoJ into the cytosol. *Traffic* **8,** 554–565.

O'Bryan, M. K., Cheema, S. S., Bartlett, P. F., Murphy, B. F., and Pearse, M. J. (1993). Clusterin levels increase during neuronal development. *J. Neurobiol.* **2,** 421–432.

Oda, T., Pasinetti, G. M., Osterburg, H. H., Anderson, C., Johnson, S. A., and Finch, C. E. (1994). Purification and characterization of brain clusterin. *Biochem. Biophys. Res. Commun.* **204,** 1131–1136.

Oda, T., Wals, P., Osterburg, H. H., Johnson, S. A., Pasinetti, G. M., Morgan, T. E., Rozovsky, I., Stine, W. B., Snyder, S. W., Holzman, T. F., Krafft, G. A., and Finch, C. E. (1995). Clusterin (apoJ) alters the aggregation of amyloid beta-peptide (Abeta 1–42) and forms slowly sedimenting A beta complexes that cause oxidative stress. *Exp. Neurol.* **136,** 22–31.

Omwancha, J., Anway, M. D., and Brown, T. R. (2009). Differential age-associated regulation of clusterin expression in prostate lobes of brown Norway rats. *Prostate* **69,** 115–125.

O'Sullivan, J., Whyte, L., Drake, J., and Tenniswood, M. (2003). Alterations in the post-translational modification and intracellular trafficking of clusterin in MCF-7 cells during apoptosis. *Cell Death Differ.* **10,** 914–927.

Ouyang, X. S., Wang, X., Lee, D. T. W., Tsao, S. W., and Wong, Y. C. (2001). Up-regulation of TRPM-2, MMP-7 and ID-1 during sex hormone-induced prostate carcinogenesis in the Noble rat. *Carcinogenesis* **22**, 965–973.

Parczyk, K., Pilarsky, C., Rachel, U., and Koch-Brandt, C. (1994). Gp80 (clusterin; TRPM-2) mRNA level is enhanced in human renal clear cell carcinomas. *J. Cancer Res. Clin. Oncol.* **120**, 186–188.

Park, Y. M., Han, M. Y., Blackburn, R. V., and Lee, Y. J. (1998). Overexpression of HSP25 reduces the level of TNF alpha-induced oxidative DNA damage biomarker, 8-hydroxy-20-deoxyguanosine, in L929 cells. *J. Cell. Physiol.* **174**, 27–34.

Park, D. C., Yeo, S. G., Wilson, M. R., Yerbury, J. J., Kwong, J., Welch, W. R., Choi, Y. K., Birrer, M. J., Mok, S. C., and Wong, K. K. (2008). Clusterin interacts with Paclitaxel and confer Paclitaxel resistance in ovarian cancer. *Neoplasia* **10**, 964–972.

Partheen, K., Levan, K., Osterberg, L., Claesson, I., Fallenius, G., Sundfeldt, K., and Horvath, G. (2008). Four potential biomarkers as prognostic factors in stage III serous ovarian adenocarcinomas. *Int. J. Cancer* **123**, 2130–2137.

Patel, N. V., Wei, M., Wong, A., Finch, C. E., and Morgan, T. E. (2004). Progressive changes in regulation of apolipoproteins E and J in glial cultures during postnatal development and aging. *Neurosci. Lett.* **371**, 199–204.

Patterson, S. G., Wei, S., Chen, X., Sallman, D. A., Gilvary, D. L., Zhong, B., Pow-Sang, J., Yeatman, T., and Djeu, J. Y. (2006). Novel role of Stat1 in the development of docetaxel resistance in prostate tumor cells. *Oncogene* **25**, 6113–6122.

Paul, C., Manero, F., Gonin, S., Kretz-Remy, C., Virot, S., and Arrigo, A. P. (2002). Hsp27 as a negative regulator of cytochrome C release. *Mol. Cell. Biol.* **22**, 816–834.

Petropoulou, C., Trougakos, I. P., Kolettas, E., Toussaint, O., and Gonos, E. S. (2001). Clusterin/apolipoprotein J is a novel biomarker of cellular senescence, that does not affect the proliferative capacity of human diploid fibroblasts. *FEBS Lett.* **509**, 287–297.

Pirkkala, L., Nykanen, P., and Sistonen, L. (2001). Roles of the heat shock transcription factors in regulation of the heat shock response and beyond. *Faseb J.* **15**, 1118–1131.

Poon, S., Easterbrook-Smith, S. B., Rybchyn, M. S., Carver, J. A., and Wilson, M. R. (2000). Clusterin is an ATP-independent chaperone with very broad substrate specificity that stabilizes stressed proteins in a folding-competent state. *Biochemistry* **39**, 15953–15960.

Poppek, D., and Grune, T. (2006). Proteasomal defense of oxidative protein modifications. *Antioxid. Redox Signal.* **8**, 173–184.

Preville, X., Salvemini, F., Giraud, S., Chaufour, S., Paul, C., Stepien, G., Ursini, M. V., and Arrigo, A. P. (1999). Mammalian small stress proteins protect against oxidative stress through their ability to increase glucose-6-phosphate dehydrogenase activity and by maintaining optimal cellular detoxifying machinery. *Exp. Cell Res.* **247**, 61–78.

Pucci, S., Bonanno, E., Pichiorri, F., Angeloni, C., and Spagnoli, L. G. (2004). Modulation of different clusterin isoforms in human colon tumorigenesis. *Oncogene* **23**, 2298–2304.

Pucci, S., Bonanno, E., Sesti, F., Mazzarelli, P., Mauriello, A., Ricci, F., Zoccai, G. B., Rulli, F., Galatà, G., and Spagnoli, L. G. (2009). Clusterin in stool: A new biomarker for colon cancer screening? *Am. J. Gastroenterol.* [Epub ahead of print].

Rastaldi, M.P., Candiano, G., Musante, L., Bruschi, M., Armelloni, S., Rimoldi, L., Tardanico, R., Sanna-Cherchi, S., Ferrario, F., Montinaro, V., Haupt, R., Parodi, S., *et al.* (2006). Glomerular clusterin is associated with PKC-alpha/beta regulation and good outcome of membranous glomerulonephritis in humans. *Kidney Int.* **70**, 477–485.

Ray, P. S., Martin, J. L., Swanson, E. A., Otani, H., Dillmann, W. H., and Das, D. K. (2001). Transgene overexpression of aB crystallin confers simultaneous protection against cardiomyocyte apoptosis and necrosis during myocardial ischemia and reperfusion. *Faseb J.* **15**, 393–402.

Reddy, K. B., Jin, G., Karode, M. C., Harmony, J. A. K., and Howe, P. H. (1996). Transforming growth factor β (TGFβ)-induced nuclear localization of Apolipoprotein J/Clusterin in epithelial cells. *Biochemistry* **35**, 6157–6163.

Redondo, M., Villar, E., Torres-Munoz, J., Tellez, T., Morell, M., and Petito, C. K. (2002). Overexpression of clusterin in human breast carcinoma. *Am. J. Pathol.* **157**, 393–399.

Renkawek, K., Voorter, C. E., Bosman, G. J., van Workum, F. P., and de Jong, W. W. (1994). Expression of aBcrystallin in Alzheimer's disease. *Acta Neuropathol. (Berl.)* **87**, 155–160.

Ricci, F., Pucci, S., Sesti, F., Missiroli, F., Cerulli, L., and Spagnoli, L. G. (2007). Modulation of Ku70/80, clusterin/ApoJ isoforms and Bax expression in indocyanine-green-mediated photooxidative cell damage. *Ophthalmic Res.* **39**, 164–173.

Rodríguez-Piñeiro, A. M., de la Cadena, M. P., López-Saco, A., and Rodríguez-Berrocal, F. J. (2006). Differential expression of serum clusterin isoforms in colorectal cancer. *Mol. Cell. Proteomics* **5**, 1647–1657.

Rosenberg, M. E., Girton, R., Finkel, D., Chmielewski, D., Barrie, A.,, III, Witte, D. P., Zhu, G., Bissler, J. J., Harmony, J. A., and Aronow, B. J. (2002). Apolipoprotein J/clusterin prevents a progressive glomerulopathy of aging. *Mol. Cell. Biol.* **22**, 1893–1902.

Saffer, H., Wahed, A., Rassidakis, G. Z., and Medeiros, L. J. (2002). Clusterin expression in malignant lymphomas. *Mod. Pathol.* **15**, 1221–1223.

Sagi, O., Wolfson, M., Utko, N., Muradian, K., and Fraifeld, V. (2005). p66ShcA and ageing: Modulation by longevity-promoting agent aurintricarboxylic acid. *Mech. Ageing Dev.* **126**, 249–254.

Sallman, D. A., Chen, X., Zhong, B., Gilvary, D. L., Zhou, J., Wei, S., and Djeu, J. Y. (2007). Clusterin mediates TRAIL resistance in prostate tumor cells. *Mol. Cancer Ther.* **6**, 2938–2947.

Santilli, G., Aronow, B. J., and Sala, A. (2003). Essential requirement of apolipoprotein J (clusterin) signaling for IkappaB expression and regulation of NF-kappaB activity. *J. Biol. Chem.* **278**, 38214–38219.

Saradha, B., Vaithinathan, S., and Mathur, P. P. (2008). Lindane alters the levels of HSP70 and clusterin in adult rat testis. *Toxicology* **243**, 116–123.

Sarkar, F. H., Li, Y., Wang, Z., and Kong, D. (2008). NF-kappaB signaling pathway and its therapeutic implications in human diseases. *Int. Rev. Immunol.* **27**, 293–319.

Saura, J., Petegnief, V., Wu, X., Liang, Y., and Paul, S. M. (2003). Microglial apolipoprotein E and astroglial apolipoprotein J expression *in vitro*: Opposite effects of lipopolysaccharide. *J. Neurochem.* **85**, 1455–1467.

Savković, V., Gantzer, H., Reiser, U., Selig, L., Gaiser, S., Sack, U., Klöppel, G., Mössner, J., Keim, V., Horn, F., and Bödeker, H. (2007). Clusterin is protective in pancreatitis through anti-apoptotic and anti-inflammatory properties. *Biochem. Biophys. Res. Commun.* **356**, 431–437.

Scaltriti, M., Brausi, M., Amorosi, A., Caporali, A., D'Arca, D., Astancolle, S., Corti, A., and Bettuzzi, S. (2004a). Clusterin (SGP-2, ApoJ) expression is downregulated in low- and high-grade human prostate cancer. *Int. J. Cancer* **108**, 23–30.

Scaltriti, M., Santamaria, A., Paciucci, R., and Bettuzzi, S. (2004b). Intracellular clusterin induces G2-M phase arrest and cell death in PC-3 prostate cancer cells1. *Cancer Res.* **64**, 6174–6182.

Schlabach, M. R., Luo, J., Solimini, N. L., Hu, G., Xu, Q., Li, M. Z., Zhao, Z., Smogorzewska, A., Sowa, M. E., Ang, X. L., Westbrook, T. F., Liang, A. C., *et al.* (2008). Cancer proliferation gene discovery through functional genomics. *Science* **319**, 620–624.

Schutte, J., Minna, J. D., and Birer, M. I. (1989). Deregulated expression of human c-jun transforms primary rat embryo cells in cooperation with an activated c-Ha-ras gene and transforms rat-1a cells as a single gene. *Proc. Natl. Acad. Sci. USA* **86**, 2257–2261.

Schwarz, M., Spath, L., Lux, C. A., Paprotka, K., Torzewski, M., Dersch, K., Koch-Brandt, C., Husmann, M., and Bhakdi, S. (2008). Potential protective role of apoprotein J (clusterin) in atherogenesis: Binding to enzymatically modified low-density lipoprotein reduces fatty acid-mediated cytotoxicity. *Thromb. Haemost.* **100**, 110–118.

Schwochau, G. B., Nath, K. A., and Rosenberg, M. E. (1998). Clusterin protects against oxidative stress *in vitro* through aggregative and nonaggregative properties. *Kidney Int.* **53**, 1647–1653.

Sciarra, A., Mariotti, G., Salciccia, S., Gomez, A. A., Monti, S., Toscano, V., and Di Silverio, F. (2008). Prostate growth and inflammation. *Steroid Biochem. Mol. Biol.* **108**, 254–260.

Sensibar, J. A., Sutkowski, D. M., Raffo, A., Buttyan, R., Griswold, M. D., Sylvester, S. R., Kozlowski, J. M., and Lee, C. (1995). Prevention of cell death induced by tumor necrosis factor alpha in LNCaP cells by overexpression of sulfated glycoprotein-2 (clusterin). *Cancer Res.* **55**, 2431–2437.

Senthil, K., Aranganathan, S., and Nalini, N. (2004). Evidence of oxidative stress in the circulation of ovarian cancer patients. *Clin. Chim. Acta* **339**, 27–32.

Shackelford, R. E., Kaufmann, W. K., and Paules, R. S. (2000). Oxidative stress and cell cycle checkpoint function. *Free Radic. Biol. Med.* **28**, 1387–1404.

Shannan, B., Seifert, M., Leskov, K., Willis, J., Boothman, D., Tilgen, W., and Reichrath, J. (2006). Challenge and promise: Roles for clusterin in pathogenesis, progression and therapy of cancer. *Cell Death Differ.* **13**, 12–19.

Sharma, A., Bhattacharya, B., Puri, R. K., and Maheshwari, R. K. (2008). Venezuelan equine encephalitis virus infection causes modulation of inflammatory and immune response genes in mouse brain. *BMC Genomics* **9**, 289.

Shaulian, E., and Karin, M. (2001). AP-1 in cell proliferation and survival. *Oncogene* **20**, 2390–2400.

Shim, Y. J., Shin, Y. J., Jeong, S. Y., Kang, S. W., Kim, B. M., Park, I. S., and Min, B. H. (2009). Epidermal growth factor receptor is involved in clusterin-induced astrocyte proliferation. *Neuroreport* **20**, 435–439.

Shinohara, H., Inaguma, Y., Goto, S., Inagaki, T., and Kato, K. (1993). aB crystallin and HSP28 are enhanced in the cerebral cortex of patients with Alzheimer's disease. *J. Neurol. Sci.* **119**, 203–208.

Singh, B. N., Rao, K. S., Ramakrishna, T., Rangaraj, N., and Rao, Ch. M. (2007). Association of alphaB-crystallin, a small heat shock protein, with actin: Role in modulating actin filament dynamics *in vivo*. *J. Mol. Biol.* **366**, 756–767.

Sintich, S. M., Steinberg, J., Kozlowski, J. M., Lee, C., Pruden, S., Sayeed, S., and Sensibar, J. A. (1999). Cytotoxic sensitivity to tumor necrosis factor-alpha in PC3 and LNCaP prostatic cancer cells is regulated by extracellular levels of SGP-2 (clusterin). *Prostate* **39**, 87–93.

Song, H., Zhang, B., Watson, M. A., Humphrey, P. A., Lim, H., and Milbrandt, J. (2009). Loss of Nkx3.1 leads to the activation of discrete downstream target genes during prostate tumorigenesis. *Oncogene* [Epub ahead of print].

Soti, C., and Csermely, P. (2003). Aging and molecular chaperones. *Exp. Gerontol.* **38**, 1037–1040.

Soti, C., Pal, C., Papp, B., and Csermely, P. (2005). Molecular chaperones as regulatory elements of cellular networks. *Curr. Opin. Cell Biol.* **17**, 210–215.

Stadtman, E. R. (1992). Protein oxidation and aging. *Science* **257**, 1220–1224.

Steinberg, J., Oyasu, R., Lang, S., Sintich, S., Rademaker, A., Lee, C., Kozlowski, J. M., and Sensibar, J. A. (1997). Intracellular levels of SGP-2 (Clusterin) correlate with tumor grade in prostate cancer. *Clin. Cancer Res.* **3**, 1707–1711.

Stocker, R., and Keaney, J. F., Jr. (2004). Role of oxidative modifications in atherosclerosis. *Physiol. Rev.* **84**, 1381–1478.

Storz, P. (2005). Reactive oxygen species in tumor progression. *Front Biosci.* **10**, 1881–1896.

Strocchi, P., Smith, M. A., Perry, G., Tamagno, E., Danni, O., Pession, A., Gaiba, A., and Dozza, B. (2006). Clusterin up-regulation following sub-lethal oxidative stress and lipid peroxidation in human neuroblastoma cells. *Neurobiol. Aging* **27**, 1588–1594.

Su, C. Y., Chong, K. Y., Edelstein, K., Lille, S., Khardori, R., and Lai, C. C. (1999). Constitutive hsp70 attenuates hydrogen peroxide-induced membrane lipid peroxidation. *Biochem. Biophys. Res. Commun.* **265**, 279–284.

Sun, Y., and MacRae, T. H. (2005). Small heat shock proteins: Molecular structure and chaperone function. *Cell. Mol. Life Sci.* **62**, 2460–2476.

Swertfegerv, D. K., Witte, D. P., Stuart, W. D., Rockman, H. A., and Harmony, J. A. (1996). Apolipoprotein J/clusterin induction in myocarditis: A localized response gene to myocardial injury. *Am. J. Pathol.* **148**, 1971–1983.

Takase, O., Minto, A. W., Puri, T. S., Cunningham, P. N., Jacob, A., Hayashi, M., and Quigg, R. J. (2008). Inhibition of NF-kappaB-dependent Bcl-xL expression by clusterin promotes albumin-induced tubular cell apoptosis. *Kidney Int.* **73**, 567–577.

Thomàs-Moyà, E., Gianotti, M., Proenza, A. M., and Lladó, I. (2006). The age-related paraoxonase 1 response is altered by long-term caloric restriction in male and female rats. *J. Lipid Res.* **47**, 2042–2048.

Thomas-Tikhonenko, A., Viard-Leveugle, I., Dews, M., Wehrli, P., Sevignani, C., Yu, D., Ricci, S., el-Deiry, W., Aronow, B., Kaya, G., Saurat, J. H., and French, L. E. (2004). Myc-transformed epithelial cells down-regulate clusterin, which inhibits their growth *in vitro* and carcinogenesis *in vivo*. *Cancer Res.* **64**, 3126–3136.

Toyokuni, S. (2008). Molecular mechanisms of oxidative stress-induced carcinogenesis: From epidemiology to oxygenomics. *IUBMB Life* **60**, 441–447.

Trougakos, I. P., and Gonos, E. S. (2002). Clusterin/Apolipoprotein J in human aging and cancer. *Int. J. Biochem. Cell Biol.* **34**, 1430–1448.

Trougakos, I. P., and Gonos, E. S. (2006). Regulation of Clusterin/Apolipoprotein J, a functional homologue to the small heat shock proteins, by oxidative stress in ageing and age-related diseases. *Free Radic. Res.* **40**, 1324–1334.

Trougakos, I. P., Chondrogianni, N., Pimenidou, A., Katsiki, M., Tzavelas, C., and Gonos, E. S. (2002a). Slowing down cellular aging *in vitro*. *In* "Modulating Aging and Longevity," (S. I. Rattan, Ed.). pp. 65–83. Kluwer Academic Publishers, The Netherlands.

Trougakos, I. P., Poulakou, M., Stathatos, M., Chalikia, A., Melidonis, A., and Gonos, E. S. (2002b). Serum levels of the senescence biomarker Clusterin/Apolipoprotein J increase significantly in diabetes type II and during development of coronary heart disease or at myocardial infarction. *Exp. Gerontol.* **37**, 1175–1187.

Trougakos, I. P., So, A., Jansen, B., Gleave, M. E., and Gonos, E. S. (2004). Silencing expression of the Clusterin/Apolipoprotein J gene in human cancer cells using small interfering RNA induces spontaneous apoptosis, reduced growth ability and cell sensitization to genotoxic and oxidative stress. *Cancer Res.* **64**, 1834–1842.

Trougakos, I. P., Lourda, M., Agiostratidou, G., Kletsas, D., and Gonos, E. S. (2005). Differential effects of Clusterin/Apolipoprotein J on cellular growth and survival. *Free Radic. Biol. Med.* **38**, 436–449.

Trougakos, I. P., Pawelec, G., Tzavelas, C., Ntouroupi, T., and Gonos, E. S. (2006a). Clusterin/Apolipoprotein J up-regulation after Zinc exposure, replicative senescence or differentiation of human haematopoietic cells. *Biogerontology* **7**, 375–382.

Trougakos, I. P., Petropoulou, C., Franceschi, C., and Gonos, E. S. (2006b). Reduced expression levels of the senescence biomarker Clusterin/Apolipoprotein J in lymphocytes from healthy centenarians. *Ann. N. Y. Acad. Sci.* **1067**, 294–300.

Trougakos, I. P., Djeu, J. Y., Gonos, E. S., and Boothman, D. A. (2009a). Advances and challenges in basic and translational research on clusterin. *Cancer Res.* **69**, 403–406.

Trougakos, I. P., Lourda, M., Antonelou, M. H., Kletsas, D., Gorgoulis, V. G., Papassideri, I. S., Zou, Y., Margaritis, L. H., Boothman, D. A., and Gonos, E. S. (2009b). Intracellular Clusterin inhibits mitochondrial apoptosis by suppressing p53-activating stress signals and stabilizing the cytosolic Ku70–Bax protein complex. *Clin. Cancer Res.* **15**, 48–59.

Tycko, B., Feng, L., Nguyen, L., Francis, A., Hays, A., Chung, W-Y., Tang, M-X., Stern, Y., Sahota, A., Hendrie, H., and Mayeux, R. (1996). Polymorphisms in the human Apolipoprotein J/Clusterin gene: Ethnic variation and distribution in Alzheimer's disease. *Hum. Genet.* **98**, 430–436.

Ungar, D. R., Hailat, N., Strahler, J. R., Kuick, R. D., Brodeur, G. M., Seeger, R. C., Reynolds, C. P., and Hanash, S. M. (1994). Hsp27 expression in neuroblastoma: Correlation with disease stage. *J. Natl Cancer Inst.* **86**, 780–784.

Upham, B. L., Sun Kang, K., Cho, H-Y., and Trosko, J. E. (1997). Hydrogen peroxide inhibits gap junctional intercellular communication in glutathione sufficient but not glutathione deficient cells. *Carcinogenesis* **18**, 37–42.

Vaithinathan, S., Saradha, B., and Mathur, P. P. (2009). Methoxychlor an organochlorine pesticide-induced alteration in the levels of HSP70 and clusterin is accompanied with oxidative stress in adult rat testis. *J. Biochem. Mol. Toxicol.* **23**, 29–35.

Valko, M., Rhodes, C. J., Moncol, J., Izakovic, M., and Mazur, M. (2006). Free radicals, metals and antioxidants in oxidative stress-induced cancer. *Chem. Biol. Interact.* **160**, 1–40.

Viard, I., Wehrli, P., Jornot, L., Bullani, R., Vechietti, J- L., Schifferli, J. A., Tschopp, J., and French, L. E. (1999). Clusterin gene expression mediates resistance to apoptotic cell death induced by heat shock and oxidative stress. *J. Invest. Dermatol.* **112**, 290–296.

Vousden, K. H., and Lane, D. P. (2007). p53 in health and disease. *Nat. Rev. Mol. Cell. Biol.* **8**, 275–283.

Wehrli, P., Charnay, Y., Vallet, P., Zhu, G., Harmony, J., Aronow, B., Tschopp, J., Bouras, C., Viard-Leveugle, I., French, L. E., and Giannakopoulos, P. (2001). Inhibition of post-ischemic brain injury by clusterin overexpression. *Nat. Med.* **7**, 977–978.

Weinberg, R. A. (2008). Mechanisms of malignant progression. *Carcinogenesis* **29**, 1092–1095.

Weitzman, S. A., Turk, P. W., Milkowski, D. H., and Kozlowski, K. (1994). Free radical adducts induce alterations in DNA cytosine methylation. *Proc. Natl. Acad. Sci. USA* **91**, 1261–1264.

Wellmann, A., Thieblemont, C., Pittaluga, S., Sakai, A., Jaffe, E. S., Siebert, P., and Raffeld, M. (2000). Detection of differentially expressed genes in lymphomas using cDNA arrays: Identification of clusterin as a new diagnostic marker for anaplastic large-cell lymphomas. *Blood* **96**, 398–404.

Wilson, M. R., and Easterbrook-Smith, S. B. (2000). Clusterin is a secreted mammalian chaperone. *Trends Biochem. Sci.* **25**, 95–98.

Wong, P., Taillefer, D., Lakins, J., Pineault, J., Chader, G., and Tenniswood, M. (1994). Molecular characterization of human TRPM-2/clusterin, a gene associated with sperm maturation, apoptosis and neurodegeneration. *Eur. J. Biochem.* **221**, 917–925.

Xie, D., Sham, J. S., Zeng, W. F., Che, L. H., Zhang, M., Wu, H. X., Lin, H. L., Wen, J. M., Lau, S. H., Hu, L., and Guan, X. Y. (2005). Oncogenic role of clusterin overexpression in multistage colorectal tumorigenesis and progression. *World J. Gastroenterol.* **11**, 3285–3289.

Yamamoto, H., Yamamoto, Y., Yamagami, K., Kume, M., Kimoto, S., Toyokuni, S., Uchida, K., Fukumoto, M., and Yamaoka, Y. (2000). Heat-shock preconditioning reduces oxidative protein denaturation and ameliorates liver injury by carbon tetrachloride in rats. *Res. Exp. Med. (Berl.)* **199**, 309–318.

Yang, C- R., Leskov, K., Hosley-Eberlein, K., Criswell, T., Pink, J. J., Kinsella, T. J., and Boothman, D. A. (2000). Nuclear clusterin/XIP8, an x-rayinduced Ku70-binding protein that signals cell death. *Proc. Natl. Acad. Sci. USA* **97**, 5907–5912.

Yeh, C-C., Hou, M-F., Tsai, S-M., Lin, S-K., Hsiao, J-K., Huang, J-C., Wang, L. H., Wu, S. H., Hou, L. A., Ma, H., and Tsai, L. Y. (2005). Superoxide anion radical, lipid peroxides and antioxidant status in the blood of patients with breast cancer. *Clin. Chim. Acta* **361**, 101–111.

Yerbury, J. J., Poon, S., Meehan, S., Thompson, B., Kumita, J. R., Dobson, C. M., and Wilson, M. R. (2007). The extracellular chaperone clusterin influences amyloid formation and toxicity by interacting with prefibrillar structures. *Faseb J.* **21**, 2312–2322.

Yoo, B. C., Kim, S. H., Cairns, N., Fountoulakis, M., and Lubec, G. (2001). Deranged expression of molecular chaperones in brains of patients with Alzheimer's disease. *Biochem. Biophys. Res. Commun.* **280**, 249–258.

Yoshizumi, M., Tsuchiya, K., and Tamaki, T. (2001). Signal transduction of reactive oxygen species and mitogen-activated protein kinases in cardiovascular disease. *J. Med. Invest.* **48**, 11–24.

Yu, B. P., and Chung, H. Y. (2006). Adaptive mechanisms to oxidative stress during aging. *Mech. Ageing Dev.* **127**, 436–443.

Yu, A. L., Fuchshofer, R., Kook, D., Kampik, A., Bloemendal, H., and Welge-Lüssen, U. (2009). Subtoxic oxidative stress induces senescence in retinal pigment epithelial cells via TGF-beta release. *Invest. Ophthalmol. Vis. Sci.* **50**, 926–935.

Zellweger, T., Kiyama, S., Chi, K., Miyake, H., Adomat, H., Skov, K., and Gleave, M. E. (2003). Overexpression of the cytoprotective protein clusterin decreases radiosensitivity in the human LNCaP prostate tumour model. *BJU Int.* **92**, 463–469.

Zhang, L. Y., Ying, W. T., Mao, Y. S., He, H. Z., Liu, Y., Wang, H. X., Liu, F., Wang, K., Zhang, D. C., Wang, Y., Wu, M., Qian, X. H., *et al.* (2003). Loss of clusterin both in serum and tissue correlates with the tumorigenesis of esophageal squamous cell carcinoma via proteomics approaches. *World J. Gastroenterol.* **9**, 650–654.

Zhang, H., Kim, J. K., Edwards, C. A., Xu, Z., Taichman, R., and Wang, C. Y. (2005). Clusterin inhibits apoptosis by interacting with activated Bax. *Nat. Cell Biol.* **7**, 909–915.

Zhou, W., Janulis, L., Park, I. I., and Lee, C. (2002). A novel anti-proliferative property of clusterin in prostate cancer cells. *Life Sci.* **72**, 11–21.

Zhu, X., Raina, A. K., Perry, G., and Smith, M. A. (2004). Alzheimer's disease: The two-hit hypothesis. *Lancet Neurol.* **3**, 219–226.

Zwain, I., and Amato, P. (2000). Clusterin protects granulosa cells from apoptotic cell death during follicular atresia. *Exp. Cell Res.* **257**, 101–110.

Index

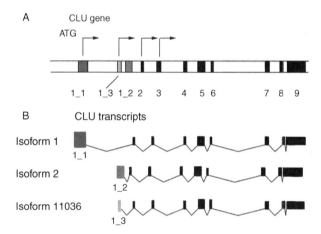

Fig. 1, Federica Rizzi *et al.*, (See Page 13 of this volume.)

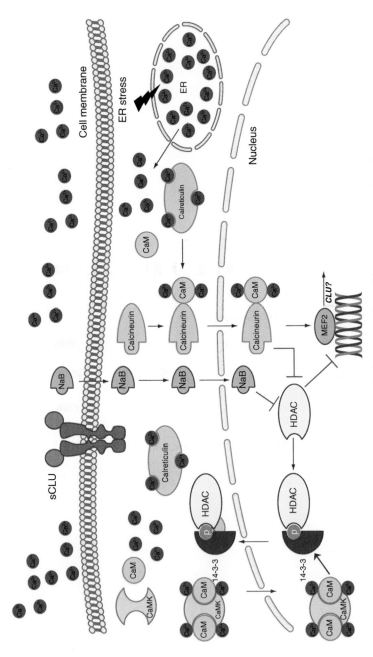

Fig. 1, Beata Pajak and Arkadiusz Orzechowski (See Page 40 of this volume.)

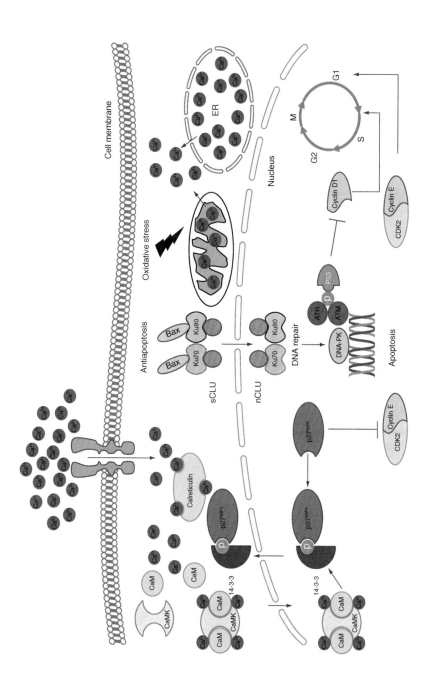

Fig. 2, Beata Pajak and Arkadiusz Orzechowski (See Page 42 of this volume.)

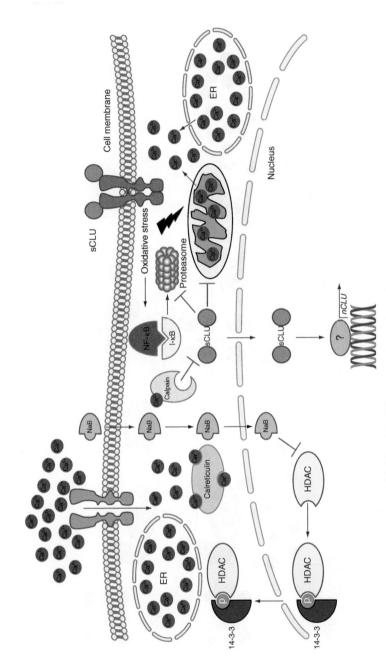

Fig. 3, Beata Pajak and Arkadiusz Orzechowski (See Page 44 of this volume.)

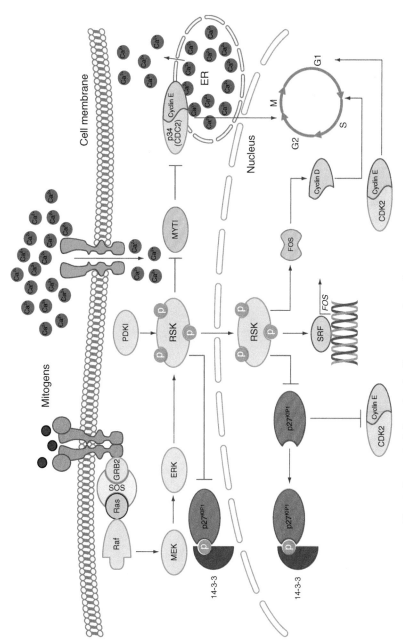

Fig. 4, Beata Pajak and Arkadiusz Orzechowski (See Page 46 of this volume.)

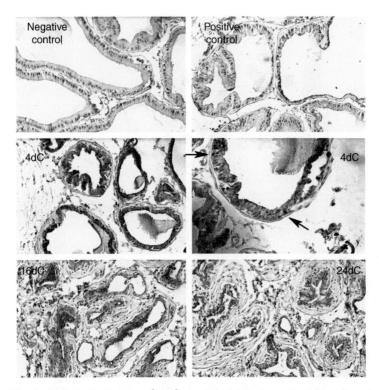

Fig. 1, Saverio Bettuzzi and Federica Rizzi (See Page 66 of this volume.)

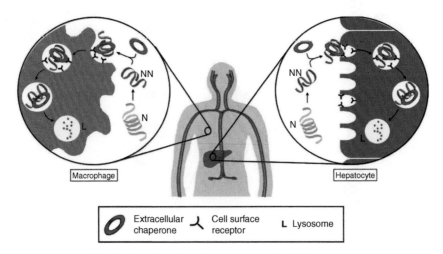

Fig. 1, Amy Wyatt *et al.*, (See Page 103 of this volume.)

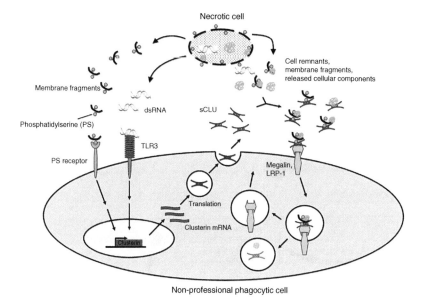

Fig. 1, Gerd Klock *et al.*, (See Page 122 of this volume.)

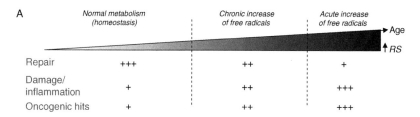

A

Normal metabolism (homeostasis) | Chronic increase of free radicals | Acute increase of free radicals

→ Age
↑ RS

Repair	+++	++	+
Damage/ inflammation	+	++	+++
Oncogenic hits	+	++	+++

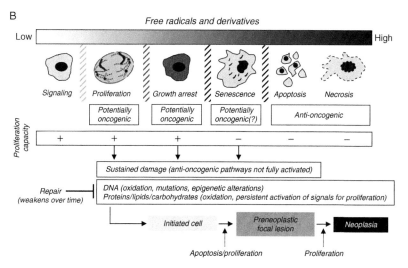

B

Free radicals and derivatives

Low   High

Signaling | Proliferation | Growth arrest | Senescence | Apoptosis | Necrosis

| Potentially oncogenic | Potentially oncogenic | Potentially oncogenic(?) | Anti-oncogenic |

Proliferation capacity

+   +   +   −   −   −

Sustained damage (anti-oncogenic pathways not fully activated)

Repair (weakens over time) — DNA (oxidation, mutations, epigenetic alterations) Proteins/lipids/carbohydrates (oxidation, persistent activation of signals for proliferation)

Initiated cell → Preneoplastic focal lesion → Neoplasia

Apoptosis/proliferation   Proliferation

Fig. 1, Ioannis P. Trougakos and Efstathios S. Gonos (See Page 175 of this volume.)

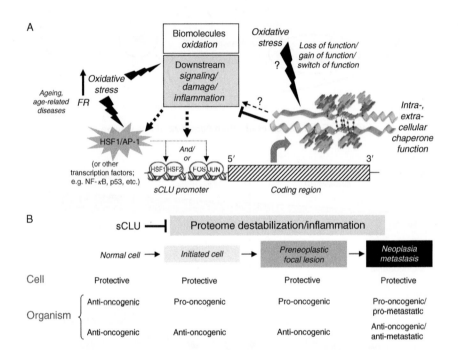

Fig. 2, Ioannis P. Trougakos and Efstathios S. Gonos (See Page 193 of this volume.)

Printed and bound by CPI Group (UK) Ltd, Croydon, CR0 4YY

08/05/2025

01864953-0001